I0019799

Merih Aydınalp Köksal

A New Approach for Modeling of Residential Energy Consumption

Merih Aydinalp Koksal

A New Approach for Modeling of Residential Energy Consumption

A National Neural Network Model for Residential End-use Energy Consumption

VDM Verlag Dr. Müller

Impressum/Imprint (nur für Deutschland/ only for Germany)
Bibliografische Information der Deutschen Nationalbibliothek: Die Deutsche Nationalbibliothek verzeichnet diese Publikation in der Deutschen Nationalbibliografie; detaillierte bibliografische Daten sind im Internet über http://dnb.d-nb.de abrufbar.
Alle in diesem Buch genannten Marken und Produktnamen unterliegen warenzeichen-, marken- oder patentrechtlichem Schutz bzw. sind Warenzeichen oder eingetragene Warenzeichen der jeweiligen Inhaber. Die Wiedergabe von Marken, Produktnamen, Gebrauchsnamen, Handelsnamen, Warenbezeichnungen u.s.w. in diesem Werk berechtigt auch ohne besondere Kennzeichnung nicht zu der Annahme, dass solche Namen im Sinne der Warenzeichen- und Markenschutzgesetzgebung als frei zu betrachten wären und daher von jedermann benutzt werden dürften.

Coverbild: www.purestockx.com

Verlag: VDM Verlag Dr. Müller Aktiengesellschaft & Co. KG
Dudweiler Landstr. 99, 66123 Saarbrücken, Deutschland
Telefon +49 681 9100-698, Telefax +49 681 9100-988, Email: info@vdm-verlag.de
Zugl.: Halifax, Dalhousie University, Diss., 2002

Herstellung in Deutschland:
Schaltungsdienst Lange o.H.G., Berlin
Books on Demand GmbH, Norderstedt
Reha GmbH, Saarbrücken
Amazon Distribution GmbH, Leipzig
ISBN: 978-3-639-01972-8

Imprint (only for USA, GB)
Bibliographic information published by the Deutsche Nationalbibliothek: The Deutsche Nationalbibliothek lists this publication in the Deutsche Nationalbibliografie; detailed bibliographic data are available in the Internet at http://dnb.d-nb.de.
Any brand names and product names mentioned in this book are subject to trademark, brand or patent protection and are trademarks or registered trademarks of their respective holders. The use of brand names, product names, common names, trade names, product descriptions etc. even without a particular marking in this works is in no way to be construed to mean that such names may be regarded as unrestricted in respect of trademark and brand protection legislation and could thus be used by anyone.

Cover image: www.purestockx.com

Publisher:
VDM Verlag Dr. Müller Aktiengesellschaft & Co. KG
Dudweiler Landstr. 99, 66123 Saarbrücken, Germany
Phone +49 681 9100-698, Fax +49 681 9100-988, Email: info@vdm-publishing.com

Printed in the U.S.A.
Printed in the U.K. by (see last page)
ISBN: 978-3-639-01972-8

To Murat, Emre, and the little bean

TABLE OF CONTENTS

ii

LIST OF TABLES

LIST OF FIGURES

LIST OF ABBREVIATIONS AND SYMBOLS

Abbreviations

A/C	Air Conditioning
ALC	Appliance, Lighting, and Space Cooling
CANN	Cascaded Neural Network
CDA	Conditional Demand Analysis
CDD	Cooling Degree Days
CI	Condition Index
CREEM	Canadian Residential Energy End-use Model
DHW	Domestic Hot Water
EM	Electricity Model
GHG	Greenhouse Gas
GLS	Generalized Least Square
HDD	Heating Degree Days
HEC	Household Energy Consumption
HRV	Heat Recovery Ventilation
MDT	Mean Daily Temperature
MLP	Multi-layer Perceptron
NGM	Natural Gas Model
NN	Neural Network
NRCan	Natural Resources of Canada
OEE	Office of Energy Efficiency
OLS	Ordinary Least Square
OM	Oil Model
RPROP	Resilient Propagation
SAE	Statistically Adjusted Engineering
SH	Space Heating
SHEU	Survey of Household Energy Use
SMM	Summer Months Method
SNNS	Stuttgart Neural Network Simulator
SSE	Sum of Square Errors
UEC	Unit Energy Consumption

Symbols

a	CDA regression coefficients
A	net heat transfer area [m^2]
ADULT	number of adults
AERATOR	number of aerators
AF	appliance features
AGECAT	dwelling construction year category
AIT	average indoor temperature [^{o}C]
AREA	heated living area [m^2]
ATTIC	dummy variable for the attic ownership
b	bias value for the hidden or output layer neurons

vii

BLANKET	dummy variable for the add-on insulation blanket around the outside of the DHW tank ownership
BP	dummy variable for the boiler pump ownership
BSMNT	dummy variable for the heated basement ownership
BWTV	dummy variable for the black and white TV ownership
c	flat spot elimination value
CAC	dummy variable for the central air condition equipment ownership
CACUSE	central air conditioning unit usage [hours/year]
CDD	cooling degree days [°C-day]
CDLOAD	clothes dryer loads [loads/week]
CHILD	number of children
CLOTH	dummy variable for the clothes washer ownership
COOK	dummy variable for the range ownership
CTV	dummy variable for the color TV ownership
CV	Coefficient of Variation
CWLOAD	clothes washer loads [loads/week]
DAYTIME	dummy variable for the daytime occupancy during weekdays
DHW	dummy variable for the domestic hot water heating equipment ownership
DISH	dummy variable for the dishwasher ownership
DOOR	number of doors
DOUBLE	number of double glazed windows
DRYER	dummy variable for the clothes dryer ownership
DTYPE	dummy variable for dwelling type
DWLOAD	dishwasher loads [loads/week]
e	error value
E	total error of the network
EDC	economic and demographic characteristics
EF	efficiency of the domestic hot water systems
EFF	efficiency of the natural gas or oil furnace/boiler [%]
FF	dummy variable for the furnace fan ownership
FLUO	number of fluorescent lamps
FREZ1	dummy variable for the main freezer ownership
FREZ2	dummy variable for the second freezer ownership
FROSTR1	dummy variable for the frost-free main refrigerator ownership
FROSTR2	dummy variable for the frost-free second refrigerator ownership
GARAGE	dummy variable for the heated garage ownership
GT	ground temperature [°C]
HALO	number of halogen lamps
HDD	heating degree days [°C-day]
HHSIZE	number of occupants in the household
HRV	dummy variable for the heat recovery ventilation system ownership
INCA	number of incandescent lamps
INCOME	household income [$10,000/yr]
LIGHTS	total number of incandescent, fluorescent, and halogen lamps
LOWFLOW	number of low-flow shower heads
m	number of neurons in the output layer, domestic hot water load [kg]
MBE	Mean Biased Error
MC	market condition

MICROW	dummy variable for the microwave ownership
N	number of end-uses, number pf patterns
n	number of neurons in the input layer
net	total input of the hidden or output layer neurons
OWNER	dummy variable for the dwelling ownership
p	number of neurons in the hidden layer, number of training patterns
PIPEINS	dummy variable for the insulation around the DHW pipes ownership
POOL	dummy variable for the pool heater ownership
POPUL	size of area of residence
PROGT	dummy variable for the programmable thermostat ownership
$\dot{q}$	heat transfer rate [W]
Q	energy consumption [GJ/yr
R	thermal resistance [m^2K/W]
R^2	fraction of variance
REF1	dummy variable for the main refrigerator ownership
REF2	dummy variable for the second refrigerator ownership
RMS	Root Mean Square
S	binary indicator of appliance ownership, partial derivative of the error function
SH	dummy variable for the space heating equipment ownership
SHAGE	age of the natural gas or oil furnace/boiler [years]
SINGLE	number of single glazed windows
SSH	dummy variable for the supplementary space heating equipment ownership
STRUC	structural features
SYSAGE	age of the DHW heating system [years]
t	target value, temperature [K]
$\bar{t}$	mean of the target value
TANK	size of the DHW tank [L]
TRIPLE	number of triple glazed windows
U	overall heat transfer coefficient [W/m^2K]
UEC	unit energy consumption [kWh/household/yr] [m^3/household/yr] [L/household/yr]
UP	appliance utilization patterns
v	weight between the input and hidden layer neurons
VCR	dummy variable for the VCR ownership
VOLF1	volume of the main freezer [L]
VOLF2	volume of the second freezer [L]
VOLR1	volume of the main refrigerator [L]
VOLR2	volume of the second refrigerator [L]
w	weight between the hidden and output layer neurons
WAC	dummy variable for the window air condition equipment ownership
WACUSE	window air conditioning unit usage [hours/year]
WC	weather condition
WINDOW	total number of single, double, and triple glazed windows
x	input to the hidden layer neuron from input layer neuron
X	matrix of the regression variables
y	output of the output layer neuron activation function
z	output of the hidden layer neuron activation function

Greek Letters

δ	error information term
η	learning parameter
ρ	maximum growth parameter, water density [kg/L]
μ	momentum term, mean of the input/output unit
ν	Quickprop weight decay term
α	RPROP increase/decrease factor
β	RPROP weight decay term
σ	standard deviation of the input/output unit
ϕ	RPROP update value
Δ	weight correction term

Subscripts

BP	boiler pump
BWTV	black and white TV
CAC	central air conditioning
CLOTH	clothes washer
COOK	range
CTV	color TV
DHW	domestic hot water heating
DISH	dishwasher
DRYER	clothes dryer
FF	furnace fan
FREZ1	main freezer
FREZ2	second freezer
i	input layer neuron, household number, indoor
j	hidden layer neuron, appliance number
k	output layer neuron
LIGHT	lighting
max	maximum value
MICROW	microwave
min	minimum value
n	scaled value of the input/output unit
o	outdoor
POOL	pool heater
REF1	main refrigerator
REF2	second refrigerator
SH	space heating
SSH	supplementary space heating
t	time period
VCR	VCR
WAC	window air conditioning

Chapter 1

Introduction

1. Background

Energy use has been a matter of policy concern since the 1970s. After the oil crises in 1973 and 1979, governments intensively promoted energy conservation. Then in the 80's, the primary focus shifted to air pollution caused by combustion of fossil fuels. In recent years, energy use and associated greenhouse gas (GHG) emissions, and their potential effects on the global climate change have been the worldwide concern.

"Greenhouse effect" can be described as the blanketing effect of the earth's atmosphere acting like the panes of glass of a greenhouse, and preventing the long wave radiation emitted by the earth's surface from dispersing into space. The principal long-lived gases responsible for absorbing outgoing radiation are carbon dioxide, methane, nitrous oxide, and chlorofluorocarbons. Among human caused emissions, carbon dioxide is by far the most globally significant GHG.

In December 1997, 160 nations gathered in Kyoto, Japan, developed the *Kyoto Protocol* that committed the developed countries to reduce GHG emissions by at least five percent below 1990 levels by 2008 - 2012. Within this legally binding agreement, Canada promised a six percent reduction below 1990 emission levels by 2010 (Environment Canada, 2001).

Improving the end-use energy efficiency is one of the most effective ways to reduce end-use energy consumption and associated emissions, especially for Canada. In 2000, the total energy consumption in Canada was about 8,200 Petajoules, making Canada one of the highest per capita energy consumers in the world (OEE, 2002). The energy consumption and associated GHG

1

emissions in each sector in 2000 are given in Figure 1.1. Mostly owing to its northerly location, and the prevalence of single family housing, close to 17% of this total, about 1,388 PJ, was for residential use, while the associated GHG emissions were 74.7 Mt, representing 16 percent of secondary energy related emissions. Thus, one of the effective means of approaching the GHG emission reductions required by the *Kyoto Protocol* is reducing the end-use energy consumption and the associated emissions from the residential sector.

Figure 1.1. Energy consumption and associated GHG emissions in Canada, sectoral distribution in 2000 (OEE, 2002)

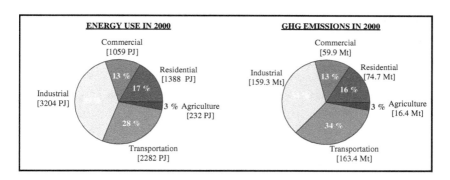

To reduce the end-use energy consumption and pollutant emissions from the residential sector, a large number of options need to be considered. These include improving the energy efficiency of dwellings through improving envelope characteristics; using higher efficiency heating equipment, household appliances and lighting; switching to less carbon-intensive fuels for space and domestic hot water heating (DHW); *etc.* Energy efficiency improvements have complex interrelated effects on the end-use energy consumption of households and the associated pollutant emissions (Ugursal and Fung, 1996; Ugursal and Fung, 1998; Farahbakhsh *et al.*, 1998). As a result, evaluating the effect of various energy efficiency improvement options on residential end-use energy consumption and associated emissions requires detailed mathematical models.

2. Problem Definition

As stated earlier, energy efficiency improvements have complex interrelated effects on the end-use energy consumption of households and the associated pollutant emissions. For example, improving the efficiency of lighting reduces the heat gain from lights, increasing the space heating energy consumption. Owing to such interrelations, detailed mathematical models are necessary to evaluate the effect of various energy efficiency improvement options on residential end-use energy consumption and associated emissions.

Recently, two approaches have been used to model residential end-use energy consumption: the engineering approach (Farahbakhsh, 1997; Farahbakhsh et al., 1997, 1998) and the conditional demand analysis (CDA) approach (Parti and Parti, 1980; Aigner et al., 1984; Fiebig et al., 1991; Kellas, 1993; Lafrange and Perron, 1994; Hsiao et al., 1995). The engineering approach involves developing a housing database representative of the national housing stock and estimating the end-use energy consumption of the households in the database using a building simulation program. A building simulation program models the end-use energy consumption of a building based on thermodynamic and physics principles taking into consideration such factors as envelope characteristics, internal and solar heat gains, weather conditions, and occupant behaviour. Thus, this approach requires a database representative of the housing stock with detailed household description data[1]. The cost of conducting a survey with sufficient information to develop input data files for a building simulation program is in the range of hundreds of dollars per household, and therefore can be prohibitive.

CDA, on the other hand, is a regression-based method. The regression essentially attributes consumption to end-uses on the basis of the household energy consumption. Since CDA does not involve modeling of the energy consumption of each household, it does not require as detailed data on the characteristics of the households as the engineering approach does; however, its results are sometimes unreliable due to multicollinearity problems (Fiebig et al., 1991; Bauwens et al., 1994; Hsiao et al., 1995). The multicollinearity problem makes it difficult to isolate the energy use of appliances with high saturation (i.e. appliances owned by a large majority of households), such as the refrigerator. Also, the model requires a very large amount of data due to the high number of

---

[1] Sufficient data are needed for each household to develop the input data file for the building simulation program used.

independent variables used in the regression equations[2].

One of the major difficulties associated with the use of the engineering approach based models to estimate the unit energy consumption (UEC) of the end-uses in the residential sector is the inclusion of socio-economic characteristics of the occupants that have a significant effect on the residential energy use. The CDA approach, which is based on regression analysis to decompose household energy consumption into appliance specific levels, can handle socio-economic factors if they are included in the model formulation.

The most comprehensive engineering approach based residential end-use energy consumption model for the Canadian housing stock is the Canadian Residential Energy End-use Model (CREEM) (Farahbakhsh, 1997; Farahbakhsh *et al.*, 1997, 1998). To develop the CREEM, data from the 1993 Survey of Household Energy Use (SHEU) (Statistics Canada, 1993) was used as the core of the database. Since the information in the 1993 SHEU database was not sufficient for this purpose, it was augmented by developing house archetypes using the Statistically Representative Housing Stock database (Ugursal and Fung, 1994; and Scanada Consultant Limited, 1992), the 1993/94 "200-House Audit" project database (NRCan, 1994), HOT2000 default values (NRCan, 1996) and minor contributions from other sources, such as engineering estimates. CREEM is a representative engineering model for the Canadian housing stock, and contains 8,767 household files. It is a versatile end-use energy model that can evaluate the impact of all types of potential energy saving measures on the residential end-use energy consumption in Canada. However, it requires extensive user expertise and lengthy input data preparation time. A detailed description of CREEM and its applications are given elsewhere (Farahbakhsh, 1997; Farahbakhsh *et al.*, 1997, 1998).

The CDA approach was used by Kellas (1993) and Lafrange and Perron (1994) to estimate the household end-use energy consumption in Manitoba and Quebec, respectively. Kellas used the data from the Residential Energy Use Survey conducted by Manitoba Hydro in 1991. The author faced difficulties in predicting the energy consumption of highly saturated appliances (*e.g.* main refrigerator) due to the multicollinearity problem. The estimates for space and DHW heating energy consumption were reasonable, but the estimates for space cooling were too high. The other Canadian CDA study was conducted by Lafrange and Perron (1994) using the data from three large surveys of Hydro-Quebec carried out between 1979 and 1989. The authors obtained reasonable

---

[2] Data on several thousands of households are needed.

4

estimates for DHW heating energy consumption, but low space heating and too high space cooling energy consumption estimates. Like Kellas (1993), the authors had difficulties to estimate the consumption of highly saturated appliances.

There is a need for an end-use energy consumption model for the Canadian housing stock that is accurate, reliable, and simpler to use than the CREEM. Such a model should be capable of characterizing the end-use energy consumption in the residential sector and should be capable of breaking down the residential energy consumption into end-use categories. The model should also be able to reflect the effects of socio-economic characteristics on residential end-use energy consumption. Such a model would be a useful tool to understand how energy is used in the residential sector, to develop optimal strategies for reducing the energy consumption[3], and evaluate the impact of energy efficiency measures on the energy consumption in the residential sector.

In this work, the Neural Network (NN) approach is used to develop a model to estimate the end-use energy consumption of Canadian single family households. Neural Networks are simplified mathematical models of biological neural networks. They are highly suitable for determining causal relationships amongst a large number of parameters such as seen in the energy consumption patterns in the residential sector. The NN approach has been used for prediction problems as a substitute for statistical approaches due to their simplicity of application and accurate estimates. A review of the published literature indicates that the NN approach has not been used or tested for national housing sector energy consumption modeling in Canada or elsewhere.

3. Objective of the Work

The major objective of this work is to develop an end-use energy consumption model of the Canadian housing stock using the NN approach. To date, efforts to model the housing sector energy consumption in Canada have been limited to the use of the engineering approach (Farahbakhsh, 1997; Farahbakhsh et al., 1997, 1998) at the national level and the CDA approach (Kellas, 1993; Lafrange and Perron 1994) at the provincial level.

---

[3] In this work, all values cited for energy consumption and savings are for "end-use" energy rather than "source" energy. This distinction is especially important in interpreting the results for electrical energy consumption and savings. "End-use" electricity consumption and savings values are to be interpreted as electricity consumption and savings at the end-user level; as such, the efficiencies of electricity generation and transmission are not reflected in these values.

The end-use energy consumption model for the Canadian housing stock, CREEM, which was developed using the engineering approach, is as advanced as possible within the constraints of the available data. Therefore, it was decided not to attempt to improve the engineering approach based model in this work.

Since none of the existing CDA models address nationwide energy consumption in the residential sector; a secondary objective of this work is to develop a new CDA model using the data from 1993 SHEU database to compare the prediction performance of the NN Model with the CDA and the Engineering Models.

Thus, the objectives of the work are as follows:

1. To develop two new energy consumption models of the Canadian residential sector representative of the Canadian housing stock, one using the NN approach and a second one using the CDA approach. Both models are to be developed based on the data from the 1993 SHEU database.

2. To estimate annual average end-use and total household energy consumption of the Canadian housing stock using the NN and CDA Models, and categorize the estimated annual average household energy consumption by dwelling type, by province, by vintage, and by space heating fuel types and energy sources;

3. To assess the accuracy of the predictions of the NN, CDA, and Engineering Models by comparing their predictions with metered data, the results from other studies, and with each other;

4. Using the NN and CDA Models developed, to assess the impact of energy efficiency measures on residential energy consumption in Canada;

5. To conduct a comparative assessment in terms of prediction performance and flexibility in evaluating the effects of the socio-economic factor and energy savings measures of the NN, CDA, and Engineering Models.

4. Scope of the Work

This work deals with the modeling of the housing sector energy consumption using the NN and CDA approaches. The models were developed using the detailed data available in the 1993 SHEU database on 8,767 households from all provinces of Canada. The 1993 SHEU was conducted by

6

Statistics Canada, and the database is representative of the Canadian housing stock. Actual energy billing data obtained from fuel suppliers and utility companies for a complete year are also available for 2,749 of the 8,767 households in the 1993 SHEU database. The households from the 1993 SHEU database with the billing data and the actual weather data for the year 1993 available from Environment Canada (1999) were used in the development of the models. Consequently, the accuracy of the models is bounded by the accuracy of the information available from these sources.

In this work, the Stuttgart Neural Network Simulator (SNNS) V4.2 software (SNNS, 1998) was used in development of the NN Model. SNNS is a well-established NN simulator that was developed at the University of Stuttgart in 1990. Since then, numerous researchers have used it as an NN development tool, including atmospheric scientists (Gottsche and Olesen, 2001; Del Frate and Schiavon, 1995), electrical and computer engineers (Binfet and Wilamowski, 2001; El-Fergany et al., 2001), medical doctors (Okon et al., 2001), and food technologists (Wittmann et al., 1997).

The SYSTAT 9.0 software (SYSTAT, 1998) was used in the development of the CDA Model. SYSTAT is a widely used statistical analysis software preferred by scientists and engineers to conduct research in areas ranging from bio-medical engineering to telecommunications, including Leach et al. (2001), Sindt et al. (2001), Bierbaum et al. (2000), Gemperline (2000), Chen (1999), Kornbrot (1999), and Miller et al. (1998).

5. Structure of the Work

The contents of the individual chapters are as follows:

Chapter 2: Review of the literature on NN and CDA methods and their uses in energy modeling, as well as overview and estimation of the models, are presented. The learning algorithms commonly used in NN modeling are also discussed.

Chapter 3: Information on the sources of data and methodologies used to develop the models to estimate the residential energy consumption using the NN and the CDA approaches are presented. The flowcharts depicting the NN and CDA methodologies are included in this chapter. The procedures used to compare the results of the models and to conduct the energy efficiency measures are also presented at the last sections of the chapter.

Chapter 4: The processes used in the development of the NN and the CDA Models are presented in detail, including the development of the datasets, input and output units, and the network architecture of the NN Model, as well as the development of the datasets and UEC equations of the CDA Model.

Chapter 5: The end-use and household energy consumption estimates obtained using the NN and the CDA Models are compared with those from the Engineering Model. The effects of socio-economic factors on the NN and the CDA Models end-use energy consumption estimates are also discussed.

Chapter 6: The impact of energy savings scenarios on DHW and space heating energy consumption estimated by the NN and CDA Models are examined and compared with those estimated by the Engineering Model.

Chapter 7: General conclusions and recommendations for future work are presented.

Chapter 2

Review of NN and CDA Modeling Approaches

1. Overview

As presented in Chapter 1, the main objective of this study is to develop two new residential energy consumption models, one using the NN approach and the second one using the CDA approach. In this chapter, a brief background of NN and CDA modeling approaches is presented, followed by a literature review on their use in energy modeling. Next, a detailed overview of each modeling approach and information on estimating energy consumption using each model are presented.

2. Review of the NN Methodology

2.1. Background

A Neural Network (NN), also commonly referred to as an Artificial Neural Network, is an information-processing model inspired by the way the densely interconnected, parallel structure of the brain processes information. In other words, neural networks are simplified mathematical models of biological neural networks. The key element of the NN is the novel structure of the information processing system. It is composed of a large number of highly interconnected processing elements that are analogous to neurons, and tied together with weighted connections that are analogous to synapses.

NNs are capable of finding internal representations of interrelations within raw data. NN are considered to be intuitive because they learn by example rather than by following programmed

9

rules. The ability to learn is one of the key aspects of NNs (Curtiss *et al.*, 1996). This typical characteristic, together with the simplicity of building and training NNs, has encouraged their application to the task of prediction. Because of their inherent non-linearity, NNs are able to identify the complex interactions between independent variables without the need for complex functional models to describe the relationships between dependent and independent variables (AlFuhaid *et al.*, 1997).

Recently, the NN approach has been proposed as a substitute for statistical approaches for classification and prediction problems. The advantages of NNs over statistical methods include the ability to classify in the presence of nonlinear relationships and the ability to perform reasonably well using incomplete databases. The comparison of the results from NNs and statistical approaches indicated that neural networks offer an accurate alternative to classical methods such as multiple regression or autoregressive models (Feuston and Thurtell, 1994; AlFuhaid *et al.*, 1997).

Although the NN concept was first introduced in 1943 (McCulloh and Pitts, 1943), it was not used extensively until the mid-1980's owing to the lack of sophisticated algorithms for general applications, and its need for fast computing resources with large storage capacity. Since the 1980's, various NN architectures and algorithms were developed (*e.g.* the multi-layer perceptron (MLP) which is generally trained with the error backpropagation algorithm, Hopfield Network, Kohonen Network, *etc.* [Hassoun, 1995]). Consequently, NN models have been used extensively as a tool for modeling, control, forecasting, and optimization in many fields of engineering and sciences such as process control, manufacturing, nuclear engineering, and pattern recognition.

2.2. Use of NNs in Energy Modeling

In the area of end-use energy consumption modeling, the application of NN has been mainly limited to utility load forecasting. There are several hundred papers in the literature on the application of NN for utility load forecasting. These clearly show the superior capability of NN models over conventional methods (such as regression analysis). Park *et al.* (1991) were the first group of researchers to use NN for load forecasting. The authors used a 3-layer[4] NN to forecast the electrical

---

[4] The definitions of the NN terms such as input, output, and hidden layers, neurons, and learning algorithms are given in Section 2.3 of Chapter 2.

load in the Seattle/Tacoma area, 1-hour and 24-hours ahead of time. Using past and current ambient temperatures and electrical load, their NN model could forecast the future load with an absolute error of about 1-2% for 1-hour, and 4% for 24-hour ahead forecasts, respectively. For 24-hour load forecasting, Peng *et al.* (1992) used an improved NN that used an alternate formulation of the problem in which the input was mapped to the output by both linear and non-linear terms, and an improved method for selecting and scaling the input units. Consequently, the absolute error in their 24-hour forecasts was less than 3% for each day of the week, with some days less than 2%.

Kiartzis *et al.* (1995) also used a 3-layer NN with 24 output neurons, one for each hour of the day (*i.e.* their model could forecast the next 24-hour load profile on an hourly basis). With a NN made up of 63 input, 70 hidden, and 24 output neurons, the yearly average absolute error of their forecasts was 2.66%. The authors expected that incorporation of additional weather information such as cloud cover, humidity, rainfall, *etc.* would further reduce the forecast error. Chen *et al.* (1996) included humidity in their NN model in addition to ambient temperature to account for the effect of humidity on air-conditioning component of the load at three types of sub-stations (residential, commercial, and industrial). The authors used a functional link network algorithm (a combination of the time series and the backpropagation algorithms) to train the network due to its higher convergence speed and accuracy. The load forecasting errors were 1.93, 2, and 2.87% for residential, commercial, and industrial substations, respectively.

AlFuhaid *et al.* (1997) used a cascaded artificial NN (CANN) to forecast half-hourly loads for the next 24-hours. The CANN approach captured the sensitivity of the non-linear influence of temperature and humidity on the load. The authors used a 3-layer NN (16 input, 8 hidden, and 3 output neurons) as the lower NN, and a 4-layer NN (107 input, 70 hidden, and 48 output neurons) as the cascaded NN. The use of the cascaded NN approach as opposed to standard NN reduced the absolute error from 3.4% to 2.7%.

NN models were used to predict energy consumption of individual buildings since they have a high potential to model nonlinear processes such as building energy loads (Kawashima, 1994). NN applications specifically to building energy analysis were pioneered by the *Joint Centre for Energy Management* at the University of Colorado, Boulder, about a decade ago. It is reported by Krarti *et al.* (1998) that Kreider and Wang (1991) were the first to apply a NN model to predict the energy consumption of a building. The electricity consumption of a commercial building was predicted and the results showed that the prediction of the NN model was accurate. The authors indicated that NN was easier to use than classical regression methods since they learn from fact

11

patterns, and there was no requirement for a *priori* statistical analysis. In a later study by the authors, the NN results were compared with statistical results for the same commercial building data (Kreider and Wang, 1992). The regression model attempted to fit all the data globally, but the accuracy at some specific points was not acceptable. The prediction of the NN model was high for those points where regression method completely missed.

Anstett and Kreider (1993) used NN to predict energy use (steam, natural gas, electricity, and water) in a complex institutional building. The authors used various network configurations, starting with a simple configuration with no hidden layers, moving progressively to more complex configurations with two or three hidden layers. The authors used the month, day of the month, day of the week, and outdoor (high, low, average) temperatures as input units, and evaluated several different training algorithms. The predictive quality of the NNs was found to be satisfactory.

In order to evaluate many of the analytical methods and to asses new methods not widely used in building data studies, an open competition was held in 1993 to identify the most accurate method for making hourly energy use predictions based on limited amount of measured data (Kreider and Haberl, 1994). More than 150 contestants requested the building data. The results of the top six models were presented in the study of Kreider and Haberl (1994). Excellent predictions were achieved by neural networks in all six models, with coefficient of variations[5] ranging from 10% to 17%. The results of the competition indicated that NN of various designs and training methods obtained more accurate values than the traditional statistical methods.

Kreider *et al.* (1995) used NN to predict the energy consumption of a complex building without knowledge of past energy consumption patterns. In this case, the forecasting problem was more difficult because the forecast was several months into the future rather than few hours. Using dry bulb temperature, humidity ratio, horizontal solar flux, wind speed, hour of the day, and weekday/weekend binary flag as inputs and recurrent (feedback) NNs (with 1- or 2-hidden layers and five or nine neurons, respectively), the authors predicted future heating and cooling loads. The authors also used the NN method to estimate the building equivalent thermal resistance and thermal capacitance from time series data on energy consumption. The assumption was that the energy consumption data contains, or implicitly represents, the characteristics of the building and its usage. Their NN model was able to estimate both the building equivalent thermal resistance and thermal

---

[5] The definition of the coefficient of variation is given in Section 2.5 of Chapter 2.

capacitance with less than 1% error.

Besides predicting building energy consumption, NN was also used to predict energy savings from building retrofits (Cohen and Krarti, 1995; Dodier and Henze, 1996). Cohen and Krarti (1995) developed a NN model from the monitored building end-use data available for a given period of time before the retrofit was implemented. Using the pre-retrofit NN model, the future building energy use without the retrofit was predicted. The energy savings were calculated from the difference between the actual post-retrofit measured data and the energy use prediction from the pre-retrofit NN model. In general, the NN model predicted savings within 10%.

Another NN approach to determine energy savings from building retrofits was proposed by Dodier and Henze (1996). The authors used one network for each energy end-use to estimate the pre-retrofit energy consumption of a commercial building. All networks had two hidden layers of 25 neurons each. The energy savings were estimated as the difference between the actual (*i.e.* measured) post-retrofit energy consumption of the building and the energy consumption predicted by the model for the pre-retrofit building. The energy savings predicted by the model obtained an average coefficient of variation of 17%.

NN approach was used by Olofsson and Andersson (2001) to predict annual energy demands of six Swedish single-family dwellings using data from daily measurements. The authors developed the NN model using daily indoor and outdoor temperature difference, as well as heating and internal use energy consumption data. The NN model had two hidden layers, each with 12 neurons, and trained with generalized delta learning algorithm[1]. The authors achieved a deviation of 4 % between the predicated and measured daily energy demands of the dwellings on an annual basis.

As this literature review indicates, NN approach has been widely used for load forecasting. Recently, the approach has been used for estimating the energy consumption of various types of commercial and residential buildings; however, NNs have not yet been used to model the residential energy consumption at the regional or national scale.

2.3. Overview of the NN Model

NNs use simple processing units, called neurons, to combine data, and store relationships between independent and dependent variables. An NN consists of several layers of neurons that are connected to each other. This connection is an information transport link from one sending to one

13

receiving neuron.

A widely used NN model called the multi-layer perceptron (MLP) NN is shown in Figure 2.1. The MLP type NN consists of one input layer, one or more hidden layers and one output layer. Each layer employs several neurons, and each neuron in a layer is connected to the neurons in the adjacent layer with different weights.

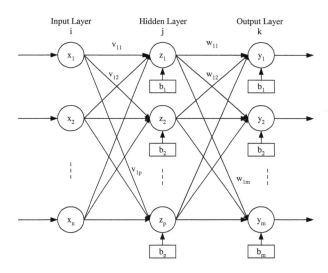

Figure 2.1. Architectural graph of an MLP with one hidden layer

Signals flow into the input layer, pass through the hidden layer(s), and arrive at the output layer. With the exception of the input layer, each neuron receives signals from the neurons of the previous layer. The incoming signals (x_{ij}) are multiplied by the weights (v_{ij}) and summed up with the bias (b_j) contribution.

$$\text{net}_j = \sum_{i=1}^{n} x_i v_{ij} + b_j \qquad (2.1)$$

where,

net_j: total input of the hidden layer neuron j

x_i: input to the hidden layer neuron j from input layer neuron i

v_{ij}: weight between the input layer neuron i and hidden layer neuron j

14

b_j: bias of the hidden layer neuron j

n: number of neurons in the input layer

The output of a neuron is determined by applying an activation function to the total input (net_j) calculated using Equation 2.1. The bias (b_j) in Equation 2.1 has the effect of increasing or decreasing the total input to the activation function, depending on whether it has a positive or negative value, respectively, and can be evaluated similar to the intercept term in a linear regression model. The bias avoids the tendency of an activation function to get "stuck" in the saturated, limiting value area of the activation function (Kreider and Wang, 1992). The bias is actually a unit connected to a neuron with a weight of one.

Activation functions for the hidden units are needed to introduce nonlinearity into the network. Without nonlinearity, hidden units would not make MLPs more powerful than just plain networks which do not have any hidden layer units, just input and output units. The sigmoid functions, such as logistic and hyperbolic tangent functions, are the most commonly used activation functions in networks trained by backpropagation (Fausett, 1994). The logistic function with the output amplitude lying inside the range [0.0 to 1.0] is shown as:

$$z_j = f(net_j) = \frac{1}{1 + e^{-net_j}} \qquad (2.2)$$

The output amplitude of the hyperbolic tangent function lies inside the range [-1.0 to 1.0], and is shown as:

$$z_j = f(net_j) = \tanh(net_j) = \frac{1 - e^{-2net_j}}{1 + e^{-2net_j}} \qquad (2.3)$$

The outputs of the activation functions become the inputs to the layers downstream. The ultimate output of a NN model, y_k, is the output of the activation function at the output layer. The activation function for the output units is mostly chosen to be logistic, hyperbolic tangent or linear (identity) functions. The identity function can be shown as

$$z_j = f(net_j) = net_j \qquad (2.4)$$

If the computed outputs do not match the known (*i.e.* target) values, NN model is in error. Then, a portion of this error is propagated backward through the network. This error is used to adjust the weight and bias of each neuron throughout the network so the next iteration error will be less for the same units. The procedure is applied continuously and repetitively for each set of inputs until there are no measurable errors, or the total error is smaller than a specified value.

At this point, the net remembers the patterns for which it was trained and is able to recognize similar patterns in new sets of data. Once the structure and training are completed, predictions from a new set of data may be done, using the already trained network. During the training process, the neural network develops the capability of recognizing different patterns and capturing relevant relationships in the training dataset[6].

Therefore, the underlying assumption in using the NN model is that the relationships between the input and the output variables in the training dataset, the testing dataset, and prediction dataset are the same. This feature of the NN Models can also be seen in CDA models, but not in Engineering Models. The models based on the engineering approach have the capability to estimate a wide range of variables, as long as the detailed house description data are available.

2.3.1. Learning Algorithms

Brief explanations of the learning algorithms used for the training of the feedforward network model are given below, including references for further reading. Somewhat more detailed descriptions are given in Appendix A.

1. Backpropagation Algorithms

(a) Standard (Plain Vanilla) Backpropagation

Vanilla backpropagation algorithm was introduced by Rumelhart and McClelland (1986) and is the most commonly used learning algorithm. During training, each output layer neuron compares its output with its target value to determine the associated error. Based on this error, the error information term is computed (derivation is given in Appendix A), which is used to distribute the error of output layer neuron back to all neurons in the hidden layer by updating the weights between the hidden layer and output layer. In a similar way, the error information term is computed for each hidden unit and used to update the weights between the hidden layer and input layer.

The weight correction term for each weight is computed by using the error information term.

---

[6] The definitions of the datasets used in the NN development are given in Section 2.4 of Chapter 2.

16

Also, a learning parameter which is usually taken as a positive number less than one, is added to the weight correction formula to reduce the changes in the weights, so that the instability (*i.e.* oscillation) of the network is prevented. The corrections to all weights are done simultaneously by adding the weight correction term to the old weight after each training pattern.

(b) Enhanced Backpropagation

This is an improved version of the standard (vanilla) backpropagation algorithm, which uses a momentum term and a flat spot elimination value. The momentum term introduces the old weight correction term as a parameter during the computation of the new weight correction term. This avoids the oscillation problem common with the standard backpropagation algorithm when the error surface has a very narrow minimum area, and keeps the weight changes going in the same direction (Anstett and Kreider, 1993).

The derivative of the activation function, which is used during the computation of the error information term, goes to zero as the unit's output approaches to the maximum and minimum values of the activation function. Thus, the flat spot elimination value is added to the derivative of the activation function to enable the network to overcome this "sticking weights" problem (SNNS, 1998). The formulae for the computation of weight correction term and error information term are given in Appendix A.

2. Quickprop

One method to speed up the learning is to use the information about the curvature of the error surface. The Quickprop algorithm is based on the assumption that the error curve can be approximated by a quadratic polynomial (*i.e.* parabola), which is concave up (Fausett, 1994). The partial derivative of the error function with respect to the given weight summed over all training patterns is referred as slope and used in the algorithm. The slope term is calculated for all weights as given in Appendix A.

The weight correction term for the hidden and output layer weights is computed by using the information about the previous weight correction term together with the previous and current value of the slope. The initial weight correction term is computed by standard (vanilla) backpropagation algorithm. The computation of the weight correction term is given in Appendix A.

3. Resilient Propagation (RPROP)

This algorithm was introduced by Riedmiller and Braun (1993). It performs a direct adaptation of the weight adjustment based on local gradient information. The main difference from other algorithms is that the adaptation is not blurred by gradient behaviour. In this algorithm, the size of the weight correction term is changed directly, *i.e.* without considering the size of the partial derivative, whose unforeseeable behaviour can disturb the adapted learning rate (Riedmiller and Braun, 1993).

Each weight is introduced by an individual update value, which determines the size of the weight correction term. The update value is computed during the learning process based on the local sign of the error function. Once the update value of each weight is adapted, the weight correction itself follows a very simple rule: if the derivative is positive (*i.e.* increasing error) the weight is decreased by its update value, if the derivative is negative, the update value is added. The formulae used for the computation of the weight update value and weight correction term are given in Appendix A.

2.4. Estimation by the NN Model

To develop an NN model, the dataset is first divided into two sets: one to be used for the training of the network, and the other for testing its performance. Approximately 70 or 80 percent of the dataset is used for training and the rest for testing (Anstett and Kreider, 1993).

After deciding which activation function to use for the hidden and output layer units, the datasets are scaled so that each value falls within the range for which the amplitude of the outputs of the chosen activation functions lie. This is done to prevent the simulated neurons from being driven too far into saturation (Highley and Hilmes, 1993), especially when the data span many orders of magnitude. Anstett and Kreider (1993) found that the [0.1 to 0.9] interval provided better results for their dataset, when logistic function and linear function were used as the activation functions for the hidden and output layer units, respectively. Kawashima (1994) scaled the input data in the [0.0 to 1.0] interval and the output data (target values) [0.1 to 0.9] interval for his network, which used logistic function for both hidden and output layer units. Thus, it is not possible to predict which activation and which scaling interval would be best suited for any given network,

18

and these should be chosen after testing various combinations.

There are no rules to establish the number of hidden layers and the number of neurons for each hidden layer for a particular application. Generally, one hidden layer is sufficient for load prediction (Kawashima, 1994; Stevenson, 1994). Network architecture is decided basically by trial and error; comparing the performance of the nets with different number of neurons in the hidden layer(s).

The weights and the biases are initialized with distributed random values before the training starts. The learning algorithm resulting in the best performance when compared to those of the other algorithms is chosen for the training of the network. The training is repeated until the sum of the square of errors (SSE) for the entire training data is less than a specified value.

A smaller number of data that have never been shown to the network during training, *i.e.* the testing dataset, is used to test the prediction performance of the network after training is complete. The output of the network for the testing dataset is then descaled to get the original units.

2.5. Assessing the Prediction Performance of NNs

To judge the prediction performance of a network, several performance measures are used. Some of these measures are given in Table 2.1, where;

t_i: target value of the ith pattern

$\bar{t}$:   mean of the target values

y_i: predicted value of the ith pattern

N: number of patterns

Table 2.1. Measures used to judge the performance of NNs

Name	Symbol	Formula	Reference
Fraction of Variance	R^2	$1 - \dfrac{\sum\limits_{i=1}^{N}(y_i - t_i)^2}{\sum\limits_{i=1}^{N} t_i^2}$	Anstett and Kreider, 1993
Root Mean Square	RMS	$\sqrt{\dfrac{\sum\limits_{i=1}^{N}(y_i - t_i)^2}{N}}$	Kreider and Wang, 1992
Coefficient of Variation	CV	$\dfrac{\sqrt{\dfrac{\sum\limits_{i=1}^{N}(y_i - t_i)^2}{N}}}{\bar{t}} \times 100$	Kreider and Haberl, 1994
Mean Bias Error	MBE	$\dfrac{\dfrac{\sum\limits_{i=1}^{N}(y_i - t_i)}{N}}{\bar{t}} \times 100$	Kawashima, 1994
Sum of Square of Errors	SSE	$\sum\limits_{i=1}^{N}(y_i - t_i)^2$	

3. Review of the CDA Methodology

3.1. Background

CDA is a regression based econometric technique designed to decompose household energy consumption into appliance-specific components. The regression breaks down the consumption into major end-uses on the basis of the association between appliance holdings and household energy consumption.

CDA was first introduced by Parti and Parti in a study conducted for *San Diego Gas & Electric Company* in 1977 (Parti and Parti, 1980). The CDA model developed by the authors separated the total household energy consumption into 16 appliance categories based on the data from 5,286 households. Their estimates of appliance energy use were reasonably close to the engineering estimates. Similarly, their estimates for price and income elasticities lied within the range of estimates presented in previous studies.

Four kinds of data are generally used to develop a CDA model: household energy consumption, generally in the form of billing records; information on the household appliance holdings and economic/demographic features, obtained from appliance saturation surveys; weather data; and information on market conditions (*e.g.* energy prices).

Besides being used to estimate the end-use energy consumption of households (with different physical and demographic characteristics), CDA is used to estimate income and price elasticities (Parti and Parti, 1980), and the hourly load of major household appliances through the day (Aigner *et al.*, 1984; Fiebeg *et al.*, 1991; Blaney *et al.*, 1994; Hsiao *et al.*, 1995). CDA can also be used, to some extent, to assess the impacts of energy conservation measures, such as increasing building envelope insulation and appliance efficiencies.

3.2. Use of CDA in Energy Modeling

Aigner *et al.* (1984) used the CDA method and 15-minute demand data from 130 households to obtain hourly end-use load profiles. The authors used 24 regression equations, each representing an hour of the day, to estimate the consumption through the day. In order to generate more precise estimates, restrictions were imposed on the parameters of the hourly equations, assuming that some appliances were not used in the early hours of the morning and hence could be excluded from those particular equations. The equations were first estimated by Ordinary Least Square (OLS) method (well-known as Gauss and Markov Theorem [Johnston and DiNardo, 1997]) and the multiple coefficient of determination values of the equations ranged approximately from 0.55 to 0.80. Generalized Least Square (GLS) (Johnston and DiNardo, 1997) was also used for the estimation of the hourly end-use loads and the results showed that GLS method estimated the hourly loads lower than the ones estimated by OLS. The estimated hourly loads for most of the appliances were reasonable, whereas some hourly load estimates of dishwasher, clothes washer, cooking range, and clothes dryer were negative (*i.e.* unreasonable).

In order to improve the accuracy of estimates, Fiebig *et al.* (1991) reformulated the standard CDA model into a random coefficient framework (Swamy, 1970). During any particular hour, the intensity of use of a particular appliance varies from household to household, and the appliance dummy variables indicate only absence or presence of the appliance and do not allow for variation in size and capacity. Therefore, the authors treated the coefficients of appliance ownership dummy variables as random variables rather than fixed variables, which also enabled the integration of metered data. The authors used a sample of 348 households, with direct metering data for two appliances, namely main tariff water heater and off peak tariff water heater which was charged at a lower than normal rate. For off peak tariff water heater, a total of 125 out of 189 households owning the appliance were directly metered, while it was 21 out of 105 for main tariff water heater. The model obtained positive hourly estimates for all appliances except the freezer. The estimated hourly loads for the off peak tariff water heater was compared with the average hourly loads obtained from the direct metering data, and the estimated hourly loads obtained from a random coefficient model which was not integrated with direct metered data. The results showed that estimates from the model integrated with metered data were closer to the direct metered data than the ones obtained from the model without the integration of the direct metered data.

Bauwens *et al.* (1994) used a Bayesian approach (Zeller, 1971) to integrate the direct metering data by viewing the data as prior information on the energy consumption of a specific appliance. A sample of 174 households from the dataset used by Fiebeg *et al.* (1991) was used to estimate end-use consumption on weekdays and weekends. Direct metering data for the main and off peak tariff water heaters were available for 21 and 87 households, respectively. The results from the standard CDA contained negative estimates for freezer and pool pump for both weekdays and weekends, whereas, the CDA model reformulated using the Bayesian approach to integrate the direct metered data obtained no negative estimates for any appliance.

To estimate the hourly electricity use of 15 residential end-uses, Blaney *et al.* (1994) used a CDA model incorporated with Bayesian priors of electricity use for each end-use estimated using the results of a previous monthly CDA and load shape representations developed from an earlier residential household metering project. The database used in the model contained information on the electricity consumption, appliance stock, physical and demographic characteristic data for 181 households and actual hourly temperature for each household's region. The estimates obtained from standard CDA were statistically significant and had the right sign, however the estimated load shapes for highly saturated appliances, *i.e.* appliances owned by the majority of the households such as clothes washer and microwave, were not reasonable. The incorporation of the Bayesian priors greatly improved the estimated load shapes for these highly saturated appliances.

The metered data information from a 1986 survey that included space heating direct metering

22

data for 30 and water heating direct metering data for 23 households were used by Hsiao *et al.* (1995) to form prior distributions of appliance consumption. Then, by applying Bayesian technique, these prior distributions were combined with the load, appliance ownership, and demographic data from a survey conducted in 1983 including 347 households. The model included the interactions of appliance dummy variables with socio-economic variables, and weather data as explanatory variables. The hourly water and space heating consumption estimates from standard conditional demand analysis approach were compared with the ones obtained from the proposed approach. The comparison done by evaluating the hourly consumption profiles showed that the proposed approach obtained more reasonable hourly estimates for water and space heating consumption.

Either random coefficient framework or Bayesian analysis were used by Fiebig *et al.* (1991), Bauwens *et al.* (1994), Blaney *et al.* (1994), and Hsiao *et al.* (1995) to integrate prior information on end-use energy consumption in terms of metered data into the CDA models to increase the accuracy and reliability of the estimates. The following studies integrated the estimates from the engineering models into their CDA models.

Caves *et al.* (1987) treated the information from the engineering approach as prior evidence on usage patterns for specific appliances, and by using Bayesian analysis, engineering estimates were integrated into CDA model to estimate hourly appliance consumption. The sample data of the analysis contained electric consumption and appliance ownership information for 129 households for two summer months in 1977. The engineering estimates were generated from a simulation program running twelve scenarios. Average loads for each of these appliances for the sample data were constructed using a weighted average of the twelve scenarios, where the weights reflected the housing type and size characteristics of the sample, and the distribution of the sample households between the weather districts. The dishwasher and central air conditioning hourly loads estimated by engineering simulation model and the proposed model were compared. For central air conditioning, both methods provided similar estimates, but for dishwasher the estimates differed considerably, since dishwasher consumption was more dependent on consumer behaviour than that of central air conditioning.

The Statistically Adjusted Engineering (SAE) model developed by Train (1992) combined engineering estimates of end-use loads with CDA. The survey data and consumption information at both the household and end-use levels of 800 households were used in the study. The engineering estimates of loads for each end-use were entered as explanatory variables in CDA model. The estimated coefficients of these variables adjusted the engineering loads statistically to reflect the

23

actual total loads of households. These estimated loads and engineering loads were then compared with metered end-use loads for the same households. It was found that SAE model improved the engineering estimates considerably for space conditioning appliances, but added error to the engineering estimates for other appliances.

One of the Canadian CDA models was developed by Kellas (1993) using the data from the Residential Energy Use Survey conducted by Manitoba Hydro in 1991 together with the 1991 weather data. The model included 38 independent variables, including appliance ownership dummy, demographic, and weather variables, and achieved multiple coefficient of determination of 0.75. The estimates of the model showed an error of 2.8% when compared with the billing data. Due to the multicollinearity problem, the author faced difficulties in predicting the energy consumption of highly saturated appliances, such as refrigerator and clothes washers. The estimates for space and DHW heating energy consumption were reasonable, but the estimates for space cooling were high, *i.e.* about 1,360 kWh/yr/household.

Lafrange and Perron (1994) used the data from 1979, 1984 and 1989 large-scale surveys for Hydro-Quebec to estimate residential end-uses using CDA approach in Quebec. The surveys included technical characteristics of the dwellings and appliances, and demographic characteristics of the occupants. The estimates for DHW heating energy consumption were reasonable, however space heating estimates were low and cooling estimates were high. The estimates showed similarities with engineering model estimates, but the space heating estimate was lower than the engineering model estimate. Like other researchers, the authors had difficulties to estimate the consumption of highly saturated appliances.

This literature review indicates that in the 80's and mid 90's the CDA approach has been widely used to model energy consumption in the residential sector. Although various methods have been used to increase the accuracy of CDA models, and the reliability of the estimates, the CDA approach has not received wide acceptance due to its low prediction performance, and the high cost of obtaining prior information in terms of metered data or engineering estimates. Like the NN approach, the capability of the CDA models is limited to the variables used in the model equation, thus limited energy efficiency measures can be tested using the CDA approach.

3.3. Overview of the CDA Model

The energy consumption of a household can be expressed as a summation of the energy consumed by each of the appliances present in the household.[7] Thus, the energy consumption of a household is directly related to the appliance stock present in the dwelling, specific features of these appliances, dwelling characteristics, and utilization patterns such as thermostat settings on water/space heaters, and behavioural patterns relating to the use of appliances.

The basic CDA model can therefore be represented in algebraic form as (EPRI, 1989):

$$HEC_{it} = \sum_{j=1}^{j} UEC_{ijt} \times S_{ij} \qquad (2.5)$$

where,

HEC_{it}: energy consumption by household i in period t

UEC_{ijt}: end-use j unit energy consumption of household i in period t

S_{ij}: a binary indicator of household i's ownership of appliance j

To develop a CDA model, the data on household energy consumption (HEC_{it}) can be obtained from utility billing records and appliance stock (S_{ij}) information can be obtained through an appliance saturation survey.

The end-use energy consumption depends upon a variety of factors and this relationship can be formalized as:

$$UEC_{ijt} = f_j \left(AF_{ij}, STRUC_i, UP_{ijt}, e_{ijt} \right) \qquad (2.6)$$

where,

AF_{ij}: features of household i's appliance j

$STRUC_i$: structural features of household i

UP_{ijt}: utilization patterns relating to appliance j

e_{ijt}: a random error term for the end-use

The effect of weather conditions (WC_{it}), market conditions (MC_{it}), and household's economic

---

[7] Here the term "appliance" is used in the most general way, including the space and domestic hot water heating as well as space cooling equipment.

25

and demographic characteristics (*EDC$_i$*) on the end-use energy utilization pattern can be shown as:

$$UP_{ijt} = g_j \left(WC_{it}, MC_{it}, EDC_i \right)$$ (2.7)

Substituting Equation 2.7 into Equation 2.6 yields:

$$UEC_{ijt} = F_j \left(AF_{ij}, STRUC_i, WC_{it}, MC_{it}, EDC_i, e_{ijt} \right)$$ (2.8)

And finally substituting Equation 2.8 into Equation 2.5 gives the general equation:

$$HEC_{it} = \sum_{j=1}^{j} F_j \left(AF_{ij}, STRUC_i, WC_{it}, MC_{it}, EDC_i, e_{ijt} \right) \times S_{ij}$$ (2.9)

Since the individual error terms are additive within their respective UEC functions, the household energy consumption equation can then be written as:

$$HEC_{it} = \sum_{j=1}^{j} F_j \left(AF_{ij}, STRUC_i, WC_{it}, MC_{it}, EDC_i \right) \times S_{ij} + e_{it}$$ (2.10)

where $e_{it} = \sum_{j=1}^{1} e_{ij} \times S_{ij}$

In most CDA models, multicollinearity problem arises, which is caused by correlation amongst the variables included in the CDA Model, limiting the capability of the regression to distinguish the impacts of these variables. Thus, the influence of some individual appliances on the total end-use energy consumption becomes difficult to separate. Mostly, appliances with high saturation cause multicollinearity problems. Moreover, it is not uncommon for this approach to yield unrealistic negative appliance consumption estimates, because of the high degree of multicollinearity.

The problem of multicollinearity is a gap between the information requirement of the model and the information provided by the sample data. The way to reduce this gap is to either expand the information content of the data, reduce the requirements of the model, or both. It is possible to reduce the problem of multicollinearity by expanding the sample size, as long as the configuration of appliance ownership is not of exactly the same pattern among individual observations. However, it requires an important expansion of sample size before a reasonable reduction in the degree of multicollinearity may be achieved (Hsiao *et al.*, 1995). Therefore, as shown in the previous section, reduction in the requirements of the model through the use of prior information in the form of data obtained by directly metering specific appliances (Fiebeg *et al.*, 1991; Bauwens *et al.*, 1994; Blaney

26

et al., 1994; Hsiao *et al.*, 1995), or engineering estimates (Caves *et al.*, 1987; Train, 1992) is an approach used by researchers with varying degrees of effectiveness.

In some cases metering data on one or more appliances is available only for a subset or a few of the total number of households, since direct metering of all of the houses in the database is not cost-effective. Thus, an appropriate method of incorporating limited direct metering data into the CDA model should be considered (Caves *et al.*, 1987; Fiebeg *et al.*, 1991; Train, 1992; Bauwens *et al.*, 1994; Blaney *et al.*, 1994; Hsiao *et al.*, 1995).

3.4. Estimation of the CDA Model

The CDA model can be estimated statistically by standard multivariate regression analysis using data on household energy consumption, appliance saturation, and other variables given in Equation 2.10.

The overall fit of the CDA model depends on the model specification and data quality. In general, the multiple coefficient of determination values of these models range from 0.55 (Aigner *et al.*, 1984) to 0.75 (Kellas, 1993). These values might seem low, but explaining the cross sectional behaviour of individual households is a difficult process since energy consumption is affected by many other factors that cannot be readily identified or quantified (tastes, habits, special circumstances), and consequently, can not be incorporated into the model. Similarly, it is not possible to incorporate all of the house characteristics (*e.g.* wall, roof, window, *etc.* areas, insulation values, infiltration, solar heat gains, climatic factors, *etc.*) into the regression model.

Once the CDA model is estimated statistically, it can be used to predict the UEC of individual households, as well as a designated group of households.

4. Closing Remarks

In this chapter, reviews of NN and CDA models used in energy consumption modeling are presented. Neural networks have been widely used in load forecasting, but there are only a few studies on commercial and residential energy consumption estimation. These few studies are limited to estimating the energy consumption of individual or a very small number of buildings.

The CDA approach has been used to disaggregate the whole house electricity consumption into end-uses. To increase the accuracy and reliability of the estimates, recent studies integrated metered data or engineering estimates into the CDA models using Bayesian or random coefficient framework techniques with varying levels of success. Despite these efforts, the CDA approach is not preferred due to the low accuracy of its estimates and the high cost of obtaining prior information in terms of metered data or engineering estimates.

The capability of the NN and the CDA approaches to estimate the impact of energy efficiency measures is limited to the input units included in the models and the dataset used to develop the models. However, the Engineering Models have the capability to estimate the impact of a wide range of measures, as long as detailed house description data are available. In the next chapter, the methodologies used to develop the NN and CDA models are presented.

Chapter 3

Research Methodology

1. Overview

The chapter begins with information on the sources of data used to develop the NN and the CDA Models. The methodologies used to develop the models are discussed in the subsequent sections. The chapter continues with the procedures used to compare the results of the models and to conduct the energy savings scenarios.

2. Sources of Data

Two sources of data were used for the development of the input units of the NN and the CDA Models: the data from the 1993 Survey of Household Energy Use (SHEU) database (Statistics Canada, 1993), and the weather and ground temperature data for 1993 (Environment Canada, 1999). The source of data for the output unit of the models was the actual energy billing data obtained from fuel suppliers and utility companies for a set of households from the 1993 SHEU.

The 1993 SHEU was conducted by Statistics Canada on behalf of Natural Resources Canada, in cooperation with the provinces of Nova Scotia, New Brunswick, Ontario, Manitoba, Saskatchewan, and with SaskPower. SHEU was commissioned to enrich the residential sector data in Canada. The target population for this survey was composed of all the housing units in Canada (excluding Yukon and North Western Territories) occupied as primary residences, both owned and rented. It is based on a mail out survey that included 376 questions. The database includes detailed information on 8,767 single-family houses from all provinces of Canada, and representative of the

29

Canadian housing sector.

The 1993 SHEU database contains detailed information on house construction, space heating/cooling and DHW heating equipment, household appliances, and socio-economic characteristics of the occupants for 8,767 households in Canada. The actual energy billing data exist for 2,749 households of the 1993 SHEU database. With the permission of the occupants, Statistics Canada obtained the complete year energy billing data of each one of these households from their fuel suppliers and utility companies.

The weather and ground temperature data for the cities where the 2,749 households are located were obtained from Environment Canada (1999). The weather data obtained includes the local mean daily temperatures (MDTs) of the households for the year 1993. The MDTs were used to calculate the heating degree days (HDD) and cooling degree days (CDD) for the locations[8].

3. NN Model

As stated previously, there are 2,749 households with energy billing data. However, only 1,067 of these households have complete energy billing data for all types of fuels and energy sources used by the households. The number of households that use electricity for DHW and space heating is 594 in the set of 1,067 households. Hence their electricity billing data account for DHW, space heating, appliance, lighting, and space cooling energy consumption. Thus, the electricity billing data of the remaining 473 households in the set of 1,067 households represents the appliance, lighting, and space cooling energy consumption, while the natural gas or oil billing data represent the DHW and space heating energy consumption. A set of 473 households is not sufficient to develop a single NN model with three output units each representing specific end-use (*i.e.* DHW heating, space heating, and appliance, lighting, and cooling). Therefore, the NN Model developed in this work consists of three networks. Each network is used to predict a single end-use energy consumption. These are:

- Space heating (SH) end-use energy consumption,
- Domestic hot water (DHW) heating end-use energy consumption,

---

[8] A base value of 18°C is taken for the HDD and CDD calculations. If the MDT is higher than 18°C, then the day is said to be a cooling day, and will have (MDT - 18) cooling degree days. If the MDT is lower than 18°C, then the day is said to be a heating day, and will have (18 - MDT) heating degree days. The annual CDD and HDD values for each city are calculated by summing the daily CDD and HDD values.

— Appliances, and lighting, and cooling (ALC) end-use energy consumption.

The development of the networks is similar and pursues the steps shown in Figure 3.1 as described below.

3.1. Development of Network Datasets

The selection of households for each network dataset was based on fuel type and energy source information, and availability of energy billing data. Fuel type and energy source information was taken from the 1993 SHEU database, and the energy billing data were obtained from Statistics Canada. Statistics.

The ALC network dataset contains 988 households with electricity bills from the 1993 SHEU database, that do not have electrical SH or DHW heating equipment. In the ALC dataset, there are 355 households that own space cooling equipment; thus, the electricity usage of these 355 households represents space cooling, appliance, and lighting energy consumption. The electricity usage of the remaining 633 households represents appliance and lighting energy consumption.

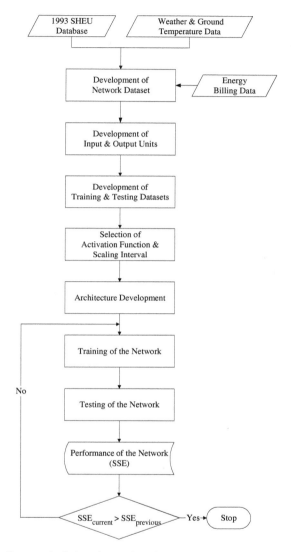

Figure 3.1. Flowchart diagram depicting the methodology used for the development of the NN Model

The number of households in the DHW dataset that use electricity or natural gas for DHW heating is 563. The dataset contains 388 households with electricity bills, and 175 with natural gas bills. Due to lack of data, households with oil bills could not be included to the DHW dataset, as explained in detail in Section 2.2.1 of Chapter 4.

The SH dataset contains 1,228 households with space heating billing data. There are 396 households with electricity billing data, 755 households with natural gas billing data, and 77 households with oil billing data for space heating, as explained in detail in Section 2.3.1 of Chapter 4.

3.2. Development of Input and Output Units

The input units of the networks describe:

- the construction details of the dwellings,
- the ownership, specifications, and usage of the space cooling equipment, appliances, and lighting,
- the specifications of the DHW and SH equipment,
- the socio-economic and behavioural characteristics of the occupants,
- the weather characteristics.

The number of input units is different for each network. Input units were selected based on their contribution to the specific end-use energy consumption. Since, the 1993 SHEU database has limited data on dwelling characteristics and heating equipment (for example, there is no detailed information on building envelope thermal characteristics), input units for these missing data were developed using information from other sources, such as HOT2000 users manual (NRCan, 1996), and studies conducted by Farahbakhsh (1997) and Farahbakhsh *et al.* (1997 and 1998). The actual energy consumption data for each household was used as the output (target) unit of the networks.

3.3. Development of the Testing and Training Datasets

The dataset of each network was divided into two sub-sets; one was used for the training of the networks and the other for the testing of the performance of the networks. As explained in Section 2.4 of Chapter 2, approximately, 70-80 percent of the dataset was used as the training set, and the remaining was used as the testing set. The households in each sub-set were selected randomly. However, special care was given to include the households with minimum and maximum input and output values into the training dataset. This is done to increase the estimation range of the NN Model.

3.4. Selection of the Activation Functions and Scaling Intervals

In this work, identity, logistic, and hyperbolic tangent functions were tested as the activation functions for the hidden and output layers. As it was pointed out in Section 2.3 of Chapter 2, these are the most commonly used activation functions for networks trained by backpropagation.

The training and testing datasets were scaled into the following intervals:

- [0.1 to 0.9]
- [-0.5 to 0.5]
- [0.0 to 1.0]
- [-1.0 to 1.0]
- [-0.9 to 0.9]

The following equations were used to scale the data into each interval:

- For the [0.1 to 0.9] interval:

$$y_n = 0.8 \left(\frac{y - y_{min}}{y_{max} - y_{min}} \right) + 0.1 \qquad (3.1)$$

- For the [-0.5 to 0.5] interval:

$$y_n = 1.0 \left(\frac{y - y_{min}}{y_{max} - y_{min}} \right) - 0.5 \qquad (3.2)$$

- For the [0.0 to 1.0] interval:

$$y_n = \left(\frac{y - y_{min}}{y_{max} - y_{min}} \right) \qquad (3.3)$$

- For the [-1.0 to 1.0] interval:

$$y_n = 2.0 \left(\frac{y - y_{min}}{y_{max} - y_{min}} \right) - 1.0 \qquad (3.4)$$

- For the [-0.9 to 0.9] interval:

$$y_n = 1.8 \left(\frac{y - y_{min}}{y_{max} - y_{min}} \right) - 0.9 \qquad (3.5)$$

where,

y_n:	value of the scaled input/output unit
y:	value of the input/output unit
y_{min}:	minimum value of the input/output unit
y_{max}:	maximum value of the input/output unit

Another method used for scaling is the normalization of the input/output units by subtracting the mean and dividing by the standard deviation:

$$y_n = \frac{y - \mu}{\sigma} \qquad (3.6)$$

where,

μ: mean of the input/output unit

σ: standard deviation of the input/output unit

Two approaches are used in this work with respect to scaling:

i) All of the data in the dataset were scaled into the intervals given above,

ii) Only continuous data in the dataset were scaled into the given intervals, and the discrete [0,1] data were left "as is", without scaling.

With five scaling intervals to be compared, a total of 11 datasets were therefore generated:

1. All data scaled to the [0.1 to 0.9] interval,
2. Only continuous data scaled to the [0.1 to 0.9] interval,
3. All data scaled to the [-0.5 to 0.5] interval,
4. Only continuous data scaled to the [-0.5 to 0.5] interval,
5. Only continuous data scaled to the [0.0 to 1.0] interval, since discrete data is already in the [0,1] range,
6. All data scaled to the [-1.0 to 1.0] interval,

35

7. Only continuous data scaled to the [-1.0 to 1.0] interval,

8. All data scaled to the [-0.9 to 0.9] interval,

9. Only continuous data scaled to the [-0.9 to 0.9] interval,

10. All data normalized,

11. Only continuous data normalized,

Identity, logistic, and hyperbolic tangent activation functions were used to develop eight networks with the configurations shown in Table 3.1. Each of the eight configurations given in Table 3.1 was tested for each of the eleven datasets that were scaled to different intervals.

Table.3.1 Network configurations tested

Network Name	Hidden Layer Activation Function	Output Layer Activation Function
Network-A	Logistic	Logistic
Network-B	Logistic	Hyperbolic Tangent
Network-C	Logistic	Identity
Network-D	Hyperbolic Tangent	Logistic
Network-E	Hyperbolic Tangent	Hyperbolic Tangent
Network-F	Hyperbolic Tangent	Identity
Network-G	Identity	Logistic
Network-H	Identity	Hyperbolic Tangent

3.5. Development of the Network Architecture

Since the number of input and output units are decided based on the available data and the desired output, respectively, only the number of units in the hidden layer(s) is left to be determined. Networks with the number of hidden layer units ranging from one to 30 or 40 were trained with each of the learning algorithm given in Section 2.3.1 of Chapter 2. The number of units in the hidden layer of the network and the learning algorithm resulting in the highest prediction performance was chosen as the network architecture for the NN Model.

After determining the number of hidden layer units and the learning algorithm resulting in the highest prediction performance, different networks with the number of hidden layer units in one, two or three layers were trained with the chosen learning algorithm to determine the best network architecture. The performance of the networks was improved by "fine-tuning" the parameters of the

36

chosen learning algorithm given in Section 2.3.1 of Chapter 2. The "fine-tuning" is done by testing a wide range of values of the learning algorithm parameters.

3.6. Training of the Networks

The training of a NN is an iterative procedure, which follows the steps described below:

(a)   Initialization of weights and biases:

All weights and biases are set to small random values between -1 and 1 (or some other suitable interval).

(b)   Feedforward propagation:

Each input unit receives an input signal and sends the signal to all units in the hidden layer. Each hidden unit sums its weighted input signals with the bias contribution, applies its activation function to compute its output signal and sends this signal to the output unit. The output unit sums its weighted input signals with the bias contribution, and applies its activation function to compute the output of the network.

(c)   Error calculation:

The output of the network, *i.e.* its prediction, and the output (target) parameter are used to compute the network error in terms of SSE. The error is used to compute the necessary changes of the weights and biases to minimize the error of the network.

(d)   Backward propagation:

The weights and biases are adjusted in a way that minimizes the error. The learning algorithms used in this work are given in Section 2.3.1 of Chapter 2. The steps from *(a)* to *(d)* are repeated until the SSE of the testing dataset stops decreasing and starts to increase, which is an indication of overtraining.

Once the networks are complete, they are used to predict the end-use energy consumption of the households in the 1993 SHEU database that are not included in the training and the testing datasets.

3.7. Assessing the Prediction Performance of the NN Model

The prediction performance of the NN Model is assessed using SSE, R^2, RMS, and CV as explained in Section 2.5 of Chapter 2.

4. CDA Model

As stated in Section 3 of Chapter 3, there are 2,749 households with energy billing data. However, there are only 1,067 households with billing data for all types of fuels and energy sources used by the households. To develop a CDA model for estimating household energy consumption, a set of 1,067 households is not sufficient, especially when there is no metered data to be used for prior information.

There are 2,050 households with electricity billing data, 1,012 households with natural gas billing data, and 236 households with oil billing data. Thus, a CDA model is developed with three components. Each component of the model was used to disaggregate the energy consumption of the households with one type of energy billing data. These models are:

- Electricity model (EM),
- Natural gas model (NGM),
- Oil model (OM).

The process used in the development of the each component of the model is similar and pursues the steps shown in Figure 3.2, and described in the following sections.

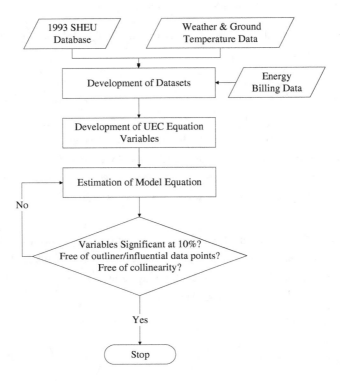

Figure 3.2. Flowchart diagram depicting the methodology used for the development of the CDA Model

4.1. Development of the Datasets

The households from the 1993 SHEU database with electricity, natural gas, or oil billing data were selected for the datasets of the CDA Model. Hence, the dataset of the Electricity CDA Model includes the households with electricity billing data, while the datasets of the natural gas and oil CDA Models include the households with natural gas and oil billing data, respectively.

4.2. Development of UEC Equations

The total energy consumption of a household is the cumulative of all energy consumed for the various end-uses in the household. As explained in Section 3.3 of Chapter 2, a unit energy consumption (UEC) equation is developed for each end-use. The 1993 SHEU database (Statistics Canada, 1993), and the 1993 weather and ground temperature data (Environment Canada, 1999) are the sources of information used in the development of the UEC equations.

The input variables of each end-use UEC equation were chosen taking into consideration the major determinants of the end-use consumption. For example, the input variables for the space heating UEC equation contain information on the structural features of the dwellings (such as heating area, type of foundation, the number of windows, doors, and floors, *etc.*), characteristics of the heating equipment, economic and demographic variables of the occupants, and weather conditions. After identifying the major determinants of the end-uses, the UEC equation for each end-use was developed as given in Equation 2.8.

The end-uses in the 1993 SHEU households are as follows:

− Electricity end-uses:
  ▪ Main and supplementary space heating,
  ▪ DHW heating,
  ▪ Space cooling,
  ▪ Major and minor appliances,
  ▪ Lighting.
− Natural gas end-uses:
  ▪ Main and supplementary space heating,
  ▪ DHW heating,
  ▪ Fireplaces,
  ▪ Cooking,
  ▪ Clothes drying,
  ▪ Pool heating.

− Oil end-uses:
  ▪ Main and supplementary space heating,
  ▪ DHW heating,
  ▪ Cooking,

40

- Pool heating.

Thus, a separate UEC equation was developed for each one of these end-uses. These are presented in Chapter 4.

4.3. Estimation of the Model Equation

As shown in detail in Chapter 4, the CDA Model equation of each component, given in Section 4 of Chapter 3, was developed combining the end-use UEC equations for each fuel type and energy source, as seen in Equation 2.10. Then, the actual energy consumption was regressed on the variables in the CDA Model equation by standard multivariate regression analysis using the SYSTAT software (SYSTAT, 1998).

4.4. Assessment of the Prediction Performance of the CDA Model

When a CDA model is first set up, it includes many variables that are expected to have an influence on the energy consumption. However, in reality, some variables have little or no influence. To test whether a variable has any influence, the student t-test is used (Johnston and DiNardo, 1997). The t-test identifies the coefficients that are not significant at a particular level. Thus, if the absolute value of the t-statistic exceeds the t-value, the hypothesis that the coefficient of a variable is zero would be rejected at a particular percent of significance. The "p-values" which are the probability of obtaining a variable as far or farther from zero (Johnston and DiNardo, 1997) are calculated by SYSTAT as part of the regression analysis. Thus, in this work, if the p-value of the variable is larger than 0.010, then the variable is not significant at 10% level, and the variable is excluded from the model equation.

As explained in Section 3.3 of Chapter 2, multicollinearity is a common problem in the CDA approach. Variables that cause multicollinearity are identified using the "condition index" (CI). To determine the CI, variables matrix (X) is multiplied by its transpose (X^T), and the eigenvalues of this matrix multiplication (X^TX) is computed. The square root of the ratio of the largest eigenvalue to the smallest eigenvalue of this matrix multiplication represents the CI of the model (Weisberg, 1985). The variables that increase the CI are the sources of multicollinearity and are excluded from the model equation.

41

The data points that do not seem to follow the same patterns as the rest of the data are called outliers, and the data points whose removal causes major changes in the results of the regression analysis are called influential (Weisberg, 1985). Outlier and influential data points are identified by tests based on computing the studentized t-values and Cook's distances (Weisberg, 1985; SYSTAT, 1998), and based on their effects on the end-use energy estimation, the data points that are identified as outlier or influential are removed one-at-a-time from the dataset.

The regression analysis is repeated by removing necessary variables from the model until all remaining variables are significant at the 10% level, there is no significant multicollinearity, and no outlier and influential data points are left. Significance level of 10% is chosen as this significance level is commonly used in estimating the end-use energy consumption in the residential sector (Lafrange and Perron, 1984; Blanet *et al.*, 1994). Once the model is complete, it can be used to predict the energy consumption of the households in the 1993 SHEU database that are not used in the model development datasets.

5. Comparative Evaluation of Models

The annual end-use and household energy consumption of the households in the 1993 SHEU database estimated using the NN and CDA Models were compared with the estimates obtained from the Engineering Model (Farahbakhsh, 1997; Farahbakhsh *et al.*, 1997, 1998), as well as with estimates reported in the literature. Comparisons were carried out for each category of households based on dwelling type (single detached and single attached) and vintage (before 1941, 1941-1960, 1960-1980, 1981-1995), province, and space heating fuel types and energy sources (electricity, oil, and natural gas).

6. Estimation of the Impact of Energy Saving Scenarios

The impact of energy saving scenarios on DHW and SH energy consumption was estimated by the NN and CDA Models, and the results were compared with those obtained from the Engineering Model. The energy saving scenarios include insulating hot water pipes, increasing the efficiencies of the DHW and SH equipment, upgrading the glazing of windows, and lowering the overnight temperature.

42

After identifying the households that could undertake energy saving scenarios, the input units of the NN Model and the variables of the CDA Model were changed to reflect the energy saving scenarios. For example, for the SH equipment efficiency upgrade scenario, the SH efficiency input unit in the SH dataset of households with standard efficiency was changed to high efficiency.

The DHW and SH energy consumption of the households that undertook the energy saving scenarios were estimated using the NN and CDA Models. The difference between the DHW and SH energy consumption estimates of the households before and after the energy efficiency upgrades provided the impact of energy saving scenarios on DHW and SH energy consumption.

7. Closing Remarks

In this chapter, methodologies used to develop the NN and the CDA Models are presented. Detailed information on the sources of data that will be used to develop the models, and procedures that will be used to assess the accuracy of the predictions of these models, as well as, procedures that will be used to evaluate the impact of various energy saving scenarios on the energy consumption in the residential sector are discussed. In the next chapter, the processes used in the development of the NN and CDA Models are described in detail.

Chapter 4

Development of the NN and CDA Models

1. Overview

This chapter presents the processes used in the development of the NN and the CDA Models. In the first section of this chapter, the development of the dataset for each network of the NN Model is presented, followed by detailed reviews of the development of the input and output units, testing and training datasets, and network architectures. In the second section of the chapter, the development of the dataset and UEC equations of each component of the CDA Model are presented.

2. Development of the NN Model

The NN model consists of three networks:

- Appliance, lighting, and space cooling (ALC) energy consumption network,
- Domestic hot water (DHW) heating energy consumption network,
- Space heating (SH) energy consumption network.

The processes used in the development of the ALC, DHW, and SH networks are presented in this section.

2.1. Appliance, Lighting, and Cooling (ALC) Network

2.1.1. Development of the ALC Network Dataset

The dataset used in the development of the ALC network is a subset of the 1993 SHEU database. There are 2,050 households in the 1993 SHEU database with electricity billing data. The electricity bills of these households account for the energy consumption for appliances, lighting, space and DHW heating (if electricity is used as the energy source), and space cooling. When these 2,050 households were categorized based on their space cooling equipment, and electric space and DHW heating equipment ownership, it was found that:

- There are 633 households that do not own space cooling equipment, or electric space and DHW heating equipment. Thus, the electricity usage of these households represents only the appliance and lighting energy consumption.

- There are 355 households that own space cooling equipment and do not use electricity for space and DHW heating equipment. Thus, the electricity usage of these households represents the space cooling, appliance, and lighting energy consumption.

Hence, the number of households with electricity billing data in the ALC network dataset is 988 (633+355). Details of the analyses of the electricity billing data are given in Appendix B.

2.1.2. Development of the ALC Network Input and Output Units

The input units of the ALC network dataset were developed using the information available from the 1993 SHEU database on appliances, lighting, and space cooling equipment of the households with electricity bills. The contribution of appliance, lighting, and space cooling energy consumption to the overall household energy consumption was considered during the selection and development of the input units.

The following sections present the final input unit dataset used for the development of the ALC NN Model. Various combinations of the input unit datasets were tested, however none produced better predictions than this presented final dataset. Some of the tested input unit datasets are given in Table 4.1.

Table 4.1. Some of the tested input unit datasets

Attempt #	Number of Input Units	Description
1	18	Seven major[9], three minor[1], five miscellaneous[1] appliances, central A/C, lighting, and area
2	30	Eight major, five minor, ten miscellaneous appliances, central and window A/C, CDD, lighting, area, income, and number of occupants
3	19	Seven major, four minor, three miscellaneous appliances, central A/C, CDD, lighting, income, and number of occupants

Input Units for Appliances

In order to reflect the contribution of the appliance energy consumption to the total household energy consumption, input units reflecting ownership, size, and usage information are included in the ALC network. The 1993 SHEU database contains information on 40 appliances, and all 40 are included in the ALC network. The number of households in the ALC network dataset that own the 40 appliances is given in Table 4.2.

---

[9] The major appliances are the refrigerators, freezers, ranges, dishwashers, clothes washers, and dryers. The minor appliances are the microwaves, TV's, VCRs, furnace fans, and boiler pumps. The miscellaneous appliances are the remaining appliances owned by the households such humidifiers, coffee makers, *etc.*

46

Table 4.2. Number and percentage of households in the ALC network dataset that own the 40 appliances (total number of households is 988)

Appliance	Number of households that own the appliance	Percentage of households that own the appliance
Main refrigerator	985	99.7%
Clothes washer	967	97.9%
Electrical cooking appliance	931	94.2%
Color TV	913	92.4%
Microwave	880	89.1%
Clothes dryer	861	87.1%
VCR	835	84.5%
Main freezer	826	83.6%
Furnace fan	741	75.0%
Stereo	668	67.6%
Kitchen exhaust fan	603	61.0%
Dishwasher	543	55.0%
Bathroom exhaust fan	521	52.7%
Car block heater	517	52.3%
Portable fan	451	45.6%
Ceiling fan	419	42.4%
CD player	342	34.6%
Second refrigerator	266	26.9%
Computer	249	25.2%
Central electronic humidifier	239	24.2%
Black and white TV	202	20.4%
Portable electric heater	202	20.4%
Central vacuum cleaner	196	19.8%
Portable humidifier	186	18.8%
Water softener	185	18.7%
Water bed	174	17.6%
Portable dehumidifier	163	16.5%
Boiler pump	137	13.9%
Sump pump	135	13.7%
Interior car warmer	118	11.9%
Electric blanket	114	11.5%
Central electronic air filter	99	10.0%
Central ventilation system	75	7.6%
Second freezer	74	7.5%
Jacuzzi	72	7.3%
Fish tank	64	6.5%
Central electronic dehumidifier	21	2.1%
Heat recovery ventilation system	19	1.9%
Water cooler	12	1.2%
Sauna	11	1.1%

Input units for appliances in the ALC network were developed from the appliance ownership, size, or usage information. A review of the ALC dataset indicated that households own only one of some appliances (*e.g.* clothes washer) and more than one of others (*e.g.* color TV). Out of the 40 appliances in the dataset, 23 appliances (main refrigerator, second refrigerator, electrical cooking appliance, dishwasher, main freezer, second freezer, clothes washer, clothes dryer, microwave, furnace fan, boiler pump, water cooler, kitchen exhaust fan, central electronic air filter, central electronic humidifier, central electronic dehumidifier, central ventilation system, heat recovery ventilation system, central vacuum cleaner, sump pump, water softener, Jacuzzi, and sauna) belong to the first category, and the remaining 17 (colour TV, black and white TV, portable electrical heater, VCR, CD player, stereo, computer, electrical blanket, water bed, portable humidifier, portable dehumidifier, car block heater, interior car warmer, fish tank, bathroom exhaust fan, ceiling fan, and portable fan) belong to the second.

To indicate the ownership of the appliances that a household owns only one of, binary variables zero and one were used as input units. Thus, the variable one was used if the household owns the appliance, and zero was used if it does not. For the appliances that the household may own more than one of, the number of units of the appliance that the household owns was used as the input unit.

Household appliance energy consumption depends on the properties and the usage pattern of the appliances. It is therefore possible to improve the prediction performance of the network by incorporating appliance properties and usage pattern information into the input units in addition to ownership.

The 1993 SHEU database, and consequently, the ALC network dataset, contains information on the size of the main and second refrigerators and freezers as shown in Tables 4.3 and 4.4, respectively. Input units reflecting the size, as well as ownership, of the main and second refrigerators are used as input units as shown in Table 4.5.

48

Table 4.3. Refrigerator size information from the 1993 SHEU database

Size	Ave. Volume	Number of Households	
		Main Refrigerator	Second Refrigerator
Bar, less than 185 L	170 L	4	29
Small, 185 – 350 L	285 L	43	82
Medium, 350 – 470 L	425 L	504	116
Large, 470 – 570 L	540 L	414	38
Very large, more than 570 L	625 L	20	1

Table 4.4. Freezer size information from the 1993 SHEU database

Size	Ave. Volume	Number of Households	
		Main Freezer	Second Freezer
Very small, less than 200 L	185 L	34	4
Small, 200 - 400 L	310 L	244	26
Medium, 400 - 510 L	450 L	323	28
Large, 510 - 650 L	595 L	182	15
Very large, more than 650 L	710 L	43	1

The 1993 SHEU database, and consequently, the ALC network dataset, contains information on the average weekly usage (in terms of number of loads) of dishwashers, clothes washers, and clothes dryers. The input units reflecting the number of loads per week, as well as ownership, of the dishwashers, clothes washers, and clothes dryers are used as input units as shown in Table 4.5.

Table 4.5. Input units used for appliances in the ALC network

Input Unit	Range
Boiler pump	0 – 1
Central electronic air filter	0 – 1
Central electronic dehumidifier	0 – 1
Central electronic humidifier	0 – 1
Central vacuum cleaner	0 – 1
Central ventilation system	0 – 1
Electrical cooking appliance	0 – 1
Furnace fan	0 – 1
Heat recovery ventilation	0 – 1
Jacuzzi	0 – 1
Kitchen exhaust fan	0 – 1
Microwave	0 – 1
Sauna	0 – 1
Sump pump	0 – 1
Water cooler	0 – 1
Water softener	0 – 1
Portable dehumidifier	0 – 2
Black and white TV	0 – 3
CD player	0 – 3
Portable electric heater	0 – 3
Portable humidifier	0 – 3
Bathroom exhaust fan	0 – 4
Computer	0 – 4
Interior car warmer	0 – 4
VCR	0 – 4
Water bed	0 – 4
Stereo	0 – 6
Car block heater	0 – 7
Ceiling fan	0 – 7
Color TV	0 – 7
Electric blanket	0 – 7
Fish tank	0 – 8
Portable fan	0 – 8
Clothes dryer [loads/week]	0 – 15
Clothes washer [loads/week]	0 – 15
Dishwasher [loads/week]	0 – 15
Main refrigerator [L]	0 – 625
Second refrigerator [L]	0 – 625
Main freezer [L]	0 – 710
Second freezer [L]	0 – 710

Input Unit for Air Conditioners

There are 275 households with central air conditioning (A/C) and 84 with window A/C units in the ALC network dataset. The information on A/C equipment is limited to the capacity and annual usage of central and window A/C equipment. Since not all of the households reported the capacity of their A/C units, it is not possible to include capacity as an input unit in the ALC network.

The information in the ALC network dataset on the annual usage of the central and window A/C units is in the form of:

1. Never
2. Only a few days
3. Less than half of the summer
4. About half of the summer
5. Most of the summer

The number of hours that the A/C units were used in 1993 was estimated by assuming that the response "about half of the summer" in the 1993 SHEU corresponds to 750 hours/yr.[10] The number of hours of usage for the other four categories were estimated based on this value. The number of households in the ALC network database with central or window A/C units, and their usage hours are given in Table 4.6.

Based on the available information in the dataset, only the annual usage of A/C equipment is used to reflect the contribution of the A/C units to the total household energy consumption. The input units that are used to represent the energy consumption of central and window A/C units are given in Table 4.7.

---

[10] 750 hours/yr is calculated as follows:
Summer months: May - August (123 days)
A/C operation: During daytime (12 h/day)
"Half of the summer": 1/2 x 123 days x 12 h/day = 738 h ~ 750 h

51

Table 4.6. Number of households with central and window A/C units and the corresponding annual usage hours

A/C Unit	Usage information in the 1993 SHEU Database	Usage Hours	Number of households
Central	Never	0 hr/yr	9
	Only a few days	0.25x750=187.5hr/yr	157
	Less than half of the summer	0.50x750=375hr/yr	67
	About half of the summer	750 hr/yr	21
	More than half of the summer	1.50x750=1125hr/yr	21
Window	Never	0 hr/yr	9
	Only a few days	0.25x750=187.5hr/yr	44
	Less than half of the summer	0.50x750=375hr/yr	22
	About half of the summer	750 hr/yr	2
	More than half of the summer	1.50x750=1,125hr/yr	7

Table 4.7. Input units used for A/C units in the ALC network

Input Unit	Range
Central A/C [hours/yr]	0 – 1,125
Window A/C [hours/yr]	0 – 1,125

Input Unit for Weather Effects

The outside temperature has an important effect on the space cooling energy consumption. Therefore, the cooling degree-day (CDD) is used as an input unit as shown in Table 4.8. For this purpose, the 1993 CDD data for the cities in which the households in the ALC network dataset are located were obtained from Environment Canada (Environment Canada, 1999).

As shown in Table 4.8, the heating degree-day (HDD) is also used as an input unit in the ALC network to reflect the temperature effect on the usage of portable electric heaters. The HDD data for the cities in the ALC network database were obtained from Environment Canada (Environment Canada, 1999).

Table 4.8. Input units used for weather effects in the ALC network

Input Unit	Range
HDD [$^{\circ}$C-day]	2,930 – 6,128
CDD [$^{\circ}$C-day]	3.7 – 405

52

Other Input Units

The lighting energy consumption is another contributor of household energy consumption. In order to reflect the effect of lighting energy consumption, the total number of halogen, fluorescent, and incandescent lights that households own are used as input units.

The other input units included in the dataset are total heated area, household income, dwelling type and ownership information, size of residence area information, number of children and adults, and ratio of employed adults in the household. These input units are given in Table 4.9. With the inclusion of these, the number of input units in the dataset increases to 55.

Output Unit

The annual electricity consumption for appliances, lighting, and space cooling is the output unit for the ALC network. Thus, the annual electricity consumption values obtained from the energy billing data for the 988 households are used as the output unit of the ALC network. The electricity consumption billing data covers the calendar year of 1993, and the consumption is given in kWh.

Table 4.9. Lighting and other input units used in the ALC network

Input Unit	Range
Halogen lights	$0 - 18$
Fluorescent lights	$0 - 46$
Incandescent lights	$0 - 106$
Total heated area [m^2]	$51.2 - 753$
Income [$1,000/yr]	$10 - 85$
Dwelling type: 1 if single-detached; 0 if single-attached	$0 - 1$
Dwelling ownership: 1 if owner; 0 if renter	$0 - 1$
Size of area of residence: 1 if population is less than 15,000 2 if population is between 15,000 and 100,000 3 if population is 100,000 or over	$1 - 3$
Employed adult ratio: Number of Employed Adults / Number of Adults	$0 - 1$
Number of children	$0 - 6$
Number of adults	$1 - 8$

2.1.3. Development of the ALC Network Training and Testing Datasets

The ALC network dataset was divided into two subsets. One of these subsets was used for training (training set) and the other was used for testing (testing set) of the network. The training set contains 741 households (75% of all households in the dataset) and the testing set contains 247 households (25% of all households in the dataset). The households in each sub-set were chosen randomly.

2.1.4. Selection of Activation Functions and Scaling Intervals

The training and the testing datasets of the ALC network were scaled to the intervals given in Section 3.4 of Chapter 3 resulting in eleven datasets. Each of the eight configurations given in Table 3.1 was tested for each of the eleven datasets that were scaled to different intervals. Thus, a total of 88 networks were tested as follows: 11 scaling intervals x 8 network configurations = 88 networks.

A network with 55 input, 25 hidden, and one output units (55:25:1) trained by standard backpropagation learning algorithm with a learning rate of 0.02 was used to compare the various scaling intervals and activation functions with respect to their prediction performance.

The training of the 55:25:1 network was halted when the testing set SSE value stopped decreasing and started to increase, which is an indication of overtraining. The prediction performance and the number of cycles for each of the eight configurations that produced the best prediction performance for each of the eleven datasets are given in Table 4.10.

As seen in Table 4.10, normalization of the data did not produce good predictions, while the prediction performance of all other scaling and activation function combinations were very good with an R^2 of 0.875 or better (except one). The network with the best prediction performance (R^2 of 0.895) used data scaled to interval [−0.5 to 0.5], the logistic function for the hidden layers, and the identity function for the output layers. Thus, in the rest of the ALC network development, the logistic function was used as the activation function for the hidden layers, the identity function was used as the activation function for the output layer, and all data in the dataset were scaled to interval [−0.5 to 0.5].

Table 4.10. Comparison of scaling intervals and activation functions

Scaling Interval	Applied to	Network Configuration	R^2	Cycles
0.1 to 0.9	All data	Network G: Identity + Logistic	0.876	318
0.1 to 0.9	Only continuous	Network G: Identity + Logistic	0.875	1093
-0.5 to 0.5	All data	Network C: Logistic + Identity	0.895	160
-0.5 to 0.5	Only continuous	Network C: Logistic + Identity	0.888	174
0.0 to 1.0	All data	Network G: Identity + Logistic	0.724	202
-1.0 to 1.0	All data	Network C: Logistic + Identity	0.888	49
-1.0 to 1.0	Only continuous	Network C: Logistic + Identity	0.890	79
-0.9 to 0.9	All data	Network C: Logistic + Identity	0.889	55
-0.9 to 0.9	Only continuous	Network C: Logistic + Identity	0.892	90
Normalization	all dataset	Network A: Logistic + Logistic	0.215	156
Normalization	Only continuous	Network A: Logistic + Logistic	0.221	87

2.1.5. Development of the ALC Network Architecture

There are 55 input data units and one output data unit in the ALC NN Model. In order to find the number of hidden layer units resulting in the best prediction performance, networks with the number of hidden layer units ranging from one to 30 were trained with the four different learning algorithms presented in Section 2.3.1 of Chapter 2. Thus, a total of 120 networks were tested as follows: 4 learning algorithms x 30 network configurations = 120 networks. The parameters of the learning algorithms used in the analysis are given in Table 4.11.

Table 4.11. Parameters of the learning algorithms used

Learning Algorithm	Parameters*
Standard Backpropagation	η : 0.010
Enhanced Backpropagation	η : 0.010, μ : 0.001, c : 0.04
Quickprop	η : 0.0015, ρ : 2.10, ν : 0.000015
Resilient Propagation	β : 1.10, $\phi_{initial}$: 0.00075, ϕ_{max} : 30

*The definitions for the parameters of the learning algorithms are given in Appendix A.

Training was halted when the testing set SSE value stopped decreasing and started to increase, *i.e.* an indication of overtraining. The prediction performance in terms of SSE, R^2, RMS, and CV of the networks with the lowest testing set SSE values amongst the 30 networks for each of the four learning algorithms is presented in Table 4.12.

As seen from Table 4.12, all learning algorithms produced very good predictions with the lowest R^2 being 0.903. The network trained with the Quickprop learning algorithm with 27 hidden layer units results in the lowest testing set SSE, RMS, and CV, and highest R^2, indicating that this network has the highest prediction performance amongst the networks tested with the four learning algorithms given in Table 4.13.

Table 4.12. Performance of networks trained using four different learning algorithms

Network	Learning Algorithm	Number of Hidden Units	SSE	R^2	RMS	CV	Number of Cycles
55:27:1	Quickprop	27	3.015	0.908	0.110	2.099	182
55:02:1	Resilient Propagation	2	3.084	0.906	0.112	2.123	90
55:02:1	Enhanced Backprop.	2	3.131	0.905	0.113	2.139	833
55:02:1	Standard Backprop.	2	3.208	0.903	0.114	2.165	1,280

To determine the network architecture that produces the best prediction performance, different network architectures with a total of 27 hidden layer units in one, two or three layers were trained with the Quickprop learning algorithm. The prediction performances of these networks are given in Table 4.13. The configuration with three hidden layers each having nine units achieved the highest prediction performance with a R^2 value of 0.909. Thus, the network with three hidden layers each having nine units (55:09:09:09:1) trained with the Quickprop learning algorithm, and using logistic function for the hidden layers, the identity function for the output layer, and dataset scaled to interval [-0.5 to 0.5] was found to be the most suitable network architecture to predict the energy consumption by household appliances, lighting, and space cooling in Canadian single family households. The architectural configuration of the ALC NN Model (55:09:09:09:1), as well as the values of the weights and biases, are given in Appendix E.

Table 4.13. Performance of networks with different architectures

Network	Number of Hidden Layer Units			SSE	R^2	RMS	CV	Number of Cycles
	Layer 1	Layer 2	Layer 3					
55:27:27:1	27	27		3.295	0.900	0.116	2.195	164
55:27:27:27:1	27	27	27	3.749	0.886	0.123	2.341	451
55:10:17:1	10	17		3.371	0.898	0.117	2.220	170
55:17:10:1	17	10		3.412	0.896	0.118	2.233	96
55:09:09:09:1	9	9	9	3.001	0.909	0.110	2.094	173

2.2. Domestic Hot Water (DHW) Network

2.2.1. Development of the DHW Network Dataset

The dataset used in the development of the DHW network is a subset of the 1993 SHEU database. It is shown in Appendix C that:

- out of the 2,050 households with electricity bills, 1,037 have electrical DHW heating systems,
- out of the 1,012 households with natural gas bills, 942 have natural gas DHW heating systems,
- out of the 236 households with oil bills, 140 have oil DHW heating systems.

As shown in Appendix C, when the energy billing data of these households were analyzed, it was found that:

- out of the 1,037 households with electrical DHW heating systems, only 388 could be included in the DHW network dataset,
- out of the 942 households with natural gas DHW heating systems, only 175 could be included in the DHW network dataset,
- out of the 140 households with oil DHW heating systems, none of them could be included in the DHW network dataset.

Hence, the number of households in the DHW network dataset is 563 (388+175).

For households with electrical DHW heating systems in the database, the annual DHW

57

electricity consumption was calculated by deducting the ALC electricity consumption estimated using the ALC NN Model from the total annual energy billing data. This introduces an error in the DHW electricity consumption data used in developing the DHW NN Model; however, this approach was used since disaggregated DHW electricity consumption data do no exist. Details of the analyses conducted to develop the DHW network dataset are given in Appendix C.

2.2.2. Development of the DHW Network Input and Output Units

The input units of the DHW network dataset were developed using the information on DHW heating system and equipment properties, DHW consumption patterns, and socio-economic characteristics of the households available from the 1993 SHEU database.

Input Units for DHW Heating System and Equipment Properties

The information available in the 1993 SHEU database on the DHW heating system and equipment properties are:

- the fuel types and energy sources used,
- number of water heaters,
- age of the system,
- size of the water tank,
- ownership information (single user or shared),
- insulation of hot water tank and hot water pipes.

Since the 1993 SHEU database does not contain information on the end-use efficiency of the DHW heating systems, the typical efficiency values given in HOT2000 (NRCan, 1996) for different types of DHW heating systems were used. These are given in Table 4.14.

Table 4.14. Typical efficiency values of the DHW heating systems (NRCan, 1996)

Energy Source or Fuel Type	End-use efficiency of the DHW Heating System	
	With Tank	Without Tank
Electricity	0.824	0.936
Oil	0.530	0.450
Natural Gas/Propane	0.554	0.460
Wood	0.300	-

All households in the DHW network dataset have DHW heating systems with tanks. Based on this information and typical end-use efficiency values given in HOT 2000 (NRCan, 1996), the end-use efficiency of the DHW heating systems of the 388 households with electricity bills were set to 0.824, the DHW heating system end-use efficiency of the 175 households with natural gas bills were set to 0.554. The data available in the DHW dataset on the distribution of DHW water tanks with respect to size and age are given in Tables 4.15 and 4.16.

Table 4.15. Water tank size information from the 1993 SHEU database

Size	Ave. Value	Number of households
Small	130 L	61
Medium	180 L	418
Large	230 L	37
Very large	280 L	47

Table 4.16. DHW heating system age information from the 1993 SHEU database

Age	Ave. Value	Number of households
1 year or less	0.5 yrs	52
2 years	2 yrs	39
3 years	3 yrs	43
4 years	4 yrs	27
5 years	5 yrs	44
6 – 7 years	6.5 yrs	41
8 – 10 years	9 yrs	112
11 – 15 years	13 yrs	120
16 years or more	18 yrs	85

The binary variables of zero and one are used as input units to indicate the presence of insulation around the hot water tank and hot water pipes, as well as the sharing of the DHW heating system by other dwellings. The variable one is used if there is insulation around the hot water tank or hot water pipes, and if the DHW heating system is shared by other dwellings, and zero is used if there is no insulation around the hot water tank or hot water pipes, and if the household is the single user of the DHW system. The input units representing the DHW heating system and equipment properties are given in Table 4.17.

59

Input Units for DHW Consumption Patterns

The 1993 SHEU database contains information on the factors affecting the DHW consumption; such as number of occupants, number of weekly clothes washer and dishwasher loads, and number of low-flow showerheads and aerators used by the household. The input units representing the DHW consumption pattern are given in Table 4.17.

Input Unit for Weather Effects

Ground temperature has an important effect on the DHW heating energy consumption. Therefore, the average annual ground temperature is used as an input unit. For this purpose, the 1993 average annual ground temperature data for the cities in which the households in the DHW network dataset are located were obtained from Environment Canada (Environment Canada, 1999).

Data on average annual ground temperatures were available for only 22 cities in 1993 from Environment Canada (Environment Canada, 1999). For those cities for which no ground temperature data exists, the data from neighbouring cities were used. The range of the ground temperature data is from 5 °C to 12 °C, as given in Table 4.17.

Other Input Units

The other input units included in the dataset are the household income, the dwelling type, dwelling ownership information, and size of residence area information. With the inclusion of these, the number of input units in the dataset increases to 18. These input units are given in Table 4.17. The size of area of residence is included into the input data set to reflect the socio-economic differences between the urban and rural population.

Table 4.17. DHW network input units

	Input Unit	Range
DHW heating system and equipment properties	End-use efficiency of the system	0.554 – 0.824
	Age of the system [years]	0.5 – 18
	Size of the water tanks [L]	130 – 280
	Number water heaters	1 – 2
	Shared with other dwellings	0 – 1
	Insulation around the water tank	0 – 1
	Insulation around hot water pipes	0 – 1
DHW consumption patterns	Number of children	0 – 5
	Number of adults	1 – 8
	Clothes washer [loads/week]	0 - 15
	Dishwasher [loads/week]	0 - 15
	Number of low-flow shower heads	0 – 3
	Number of aerators	0 – 4
Weather effects	Ground temperature [$^{\circ}$C]	5 - 12
Socio-economic characteristics of the households	Income [$1,000/yr]	10 – 85
	Dwelling type: 1 if single-detached; 0 if single-attached	0 – 1
	Dwelling ownership: 1 if owner; 0 if renter	0 – 1
	Size of area of residence: 1 if population is less than 15,000 2 if population is between 15,000 and 100,000 3 if population is 100,000 or over	1 – 3

Output Unit

The annual DHW heating electricity and natural gas consumption is the output unit of the DHW network. Thus, the annual electricity and natural gas consumption values obtained from the energy billing data for the 563 households are used as the output unit of the DHW network. The electricity and natural gas consumption data cover the calendar year of 1993, and were converted to GJ.

2.2.3. Development of the DHW Network Training and Testing Datasets

The DHW network dataset was divided into two subsets. One of these subsets was used for training (training set) and the other was used for testing (testing set) of the network. The training set contains 422 households (75% of all households in the dataset) and the testing set contains 141 households (25% of all households in the dataset). The households in each sub-set were chosen randomly.

2.2.4. Selection of Activation Functions and Scaling Intervals

The training and the testing datasets of the DHW heating network were scaled to the intervals given in Section 3.4 of Chapter 3 resulting in eleven datasets. Each of the eight configurations given in Table 3.1 was tested for each of the eleven datasets that were scaled to different intervals. Thus, a total of 88 networks were tested (= 11 scaling intervals x 8 network configurations).

A network with 18 input, 10 hidden, and one output units (18:10:1) trained by standard backpropagation learning algorithm with a learning rate of 0.02 was used to compare the various scaling intervals and activation functions with respect to their prediction performance.

The training of the 18:10:1 network was halted when the testing set SSE value stopped decreasing and started to increase, which is an indication of overtraining. The prediction performance and the number of cycles for each of the eight configurations that produced the best prediction performance for each of the eleven datasets are given in Table 4.18.

As seen in Table 4.18, normalization of the data did not produce good predictions, which was also seen during the development of the ALC network. The other scaling and activation function combinations resulted in predictions with R^2 higher than 0.590. The network with the best prediction performance (R^2 of 0.863) used the dataset with only continuous data scaled to interval [0.1 to 0.9], and the logistic function for the hidden and output layers. Thus, in the rest of the DHW network development, the logistic function was used as the activation function for the hidden and output layer, and only continuous data in the sets were scaled to interval [0.1 to 0.9].

Table 4.18. Comparison of scaling intervals and activation functions

Scaling Interval	Applied to	Network Configuration	R^2	Cycles
0.1 to 0.9	All data	Network A: Logistic + Logistic	0.862	438
0.1 to 0.9	Only continuous	Network A: Logistic + Logistic	0.863	442
-0.5 to 0.5	All data	Network B: Logistic + TanH	0.595	57
-0.5 to 0.5	Only continuous	Network B: Logistic + TanH	0.593	69
0.0 to 1.0	All data	Network A: Logistic + Logistic	0.766	315
-1.0 to 1.0	All data	Network H: Identity + TanH	0.608	191
-1.0 to 1.0	Only continuous	Network H: Identity + TanH	0.607	186
-0.9 to 0.9	All data	Network H: Identity + TanH	0.607	258
-0.9 to 0.9	Only continuous	Network H: Identity + TanH	0.607	244
Normalized	All dataset	Network D: TanH + Logistic	0.179	16
Normalized	Only continuous	Network B: Logistic + TanH	0.175	54

2.2.5. Development of the DHW Network Architecture

There are 18 input data units and one output data unit in the DHW NN Model. In order to find the number of hidden layer units resulting in the best prediction performance, networks with the number of hidden layer units ranging from one to 40 were trained with the four different learning algorithms presented in Section 2.3.1 of Chapter 2. Thus, a total of 160 networks were tested as follows: 4 learning algorithms x 40 network configurations = 160 networks. The parameters of the learning algorithms used in the analysis are given in Table 4.19.

Table 4.19. Parameters of the learning algorithms used

Learning Algorithm	Parameters*
Standard Backpropagation	$\eta : 0.025$
Enhanced Backpropagation	$\eta : 0.015, \mu : 0.1, c : 0.1$
Quickprop	$\eta : 0.005, \rho : 1.5, \nu : 0.0005$
Resilient Propagation	$\beta : 1.7, \phi_{initial}: 0.02, \phi_{max}: 30$

*The definitions for the parameters of the learning algorithms are given in Appendix A.

Training was halted when the testing set SSE value stopped decreasing and started to increase, which is an indication of overtraining. The prediction performance in terms of SSE, R^2, RMS, and CV of the networks with the lowest testing set SSE values amongst the 40 networks for each of the four learning algorithms is presented in Table 4.20.

As seen from Table 4.20, the learning algorithms produced good predictions within the range of R^2 of 0.869 to 0.871. The network trained with the Resilient Propagation learning algorithm with 29 hidden layer units resulted in the lowest testing set SSE, RMS, and CV, and highest R^2, indicating that this network has the highest prediction performance amongst the networks tested with the four learning algorithms given in Table 4.19.

Table 4.20. Performance of networks trained using four different learning algorithms

Network	Learning Algorithm	Number of Hidden Units	SSE	R^2	RMS	CV	Number of Cycles
18:29:1	Resilient Propagation	29	2.518	0.871	0.134	3.337	28
18:29:1	Quickprop	29	2.521	0.871	0.134	3.340	21
18:29:1	Enhanced Backprop.	29	2.549	0.869	0.134	3.358	45
18:29:1	Standard Backprop.	29	2.551	0.869	0.135	3.360	57

To determine the network architecture that produces the best prediction performance, different network architectures with a total of 29 hidden layer units in one, two or three layers were trained with the Resilient Propagation learning algorithm. The prediction performances of these networks are given in Table 4.21. None of these configurations resulted in better prediction performance than the network with 29 hidden units in one hidden layer. Thus, the network with one hidden layer with 29 units (18:29:1) trained with the Resilient Propagation learning algorithm, and using logistic function for the hidden and output layers and dataset scaled to interval [0.1 to 0.9] was found to be the most suitable network architecture to predict the DHW heating energy consumption in Canadian single family households. The architectural configuration of the DHW NN Model (18:29:1), as well as the values of the weights and biases, are given in Appendix E.

Table 4.21. Performance of networks with different architectures

Network	Number of Hidden Layer Units			SSE	R^2	RMS	CV	Number of Cycles
	Layer 1	Layer 2	Layer 3					
18:29:1	29			2.518	0.871	0.134	3.337	28
18:29:29:1	29	29		2.645	0.864	0.137	3.421	38
18:29:29:29:1	29	29	29	2.687	0.862	0.138	3.448	25
18:14:15:1	14	15		2.635	0.865	0.137	3.415	28
18:15:14:1	15	14		2.614	0.866	0.136	3.401	30
18:10:19:1	10	19		2.676	0.863	0.138	3.441	19
18:19:10:1	19	10		2.605	0.867	0.136	3.395	31
18:9:10:10:1	9	10	10	2.650	0.864	0.137	3.424	35
18:10:9:10:1	10	9	10	2.585	0.868	0.135	3.382	30
18:10:10:9:1	10	10	9	2.665	0.864	0.137	3.433	28

2.3. Space Heating (SH) Network

2.3.1. Development of the SH Network Dataset

The dataset used in the development of the SH network is a subset of the 1993 SHEU database. It is shown in Appendix D that:

- out of the 2,050 households with electricity bills, 556 have electrical SH systems,
- out of the 1,012 households with natural gas bills, 978 have natural gas SH systems,
- out of the 236 households with oil bills, 227 have oil SH systems.

As shown in Appendix D, when the energy billing data of these households were analyzed, it was found that:

- out of the 556 households with electrical SH systems, only 396 could be included in the SH network dataset,
- out of the 978 households with natural gas SH systems, only 755 could be included in the SH network dataset,
- out of the 227 households with oil SH systems, only 77 could be included in the SH network dataset.
- Hence, the number of households in the SH network dataset is 1,228 (396+755+77).

For electrically heated households in the database, the annual SH electricity consumption was calculated by deducting the ALC and DHW heating electricity consumption estimated using the

65

ALC and DHW NN Models from the total annual energy billing data. For natural gas heated households in the database, the annual SH natural gas consumption was calculated by deducting the DHW heating natural gas consumption estimated by the DHW NN Model from the total annual energy billing data. This introduces an error in the SH energy consumption data used in developing the SH NN Model; however, this approach was used since disaggregated SH energy consumption data do no exist.

The households with oil fuelled space and DHW heating systems could not be included in the SH NN database, because it was not possible to estimate their DHW heating energy consumption using the DHW NN model due to lack of data. The households using oil only for SH were included in the SH NN dataset. Details of the analyses conducted to develop the SH network dataset are given in Appendix D.

2.3.2. Development of the SH Network Input and Output Units

SH energy consumption of a dwelling can be determined by conducting an energy balance between heat losses and heat gains as follows:

$$\text{SH Energy Consumption} = \sum \text{Heat Losses} - \sum \text{Heat Gains} \qquad (4.1)$$

where,

$\sum \text{Heat Losses} =$ Heat losses due to transmission through the building envelope

+ Heat loss due to infiltration and ventilation

$\sum \text{Heat Gains} =$ Internal heat gain from people, lighting, and appliances + Solar heat gain

The magnitude of the heat losses and gains depend on the building envelope thermal characteristics and area, envelope tightness, outdoor weather conditions, occupant behaviour, presence of electric appliances and lighting, and solar radiation. Detailed information on these factors is given below.

– Building envelope thermal characteristics and exposed area

The heat transferred through walls, ceiling, roof, widows, floors, and doors is estimated from

$$\dot{q} = UA(t_i - t_o) \tag{4.2}$$

where $\dot{q}$ is the heat transfer rate (W), U is the overall heat transfer coefficient (W/m²K), A (m²) is the net area for the given component for which U was calculated, and t_i and t_o (K) are the indoor and outdoor temperatures, respectively. The U-value of a component of a dwelling is determined by summing the resistances of the materials and surfaces involved, and then inverting this total resistance to give a transmittance or conductivity value.

– Building envelope tightness

One of the major heat losses in the dwelling is the heat required to warm outdoor air entering the dwelling through cracks and crevices around doors, windows, lighting fixtures, and joints between walls and floor, and through the building material itself.

– Internal and solar heat gains

Internal heat gains are from people, lighting, and appliances, whereas solar heat gain is a result of solar radiation entering a building through opaque and transparent surfaces such as fenestration, skylights, *etc*. Solar hear gain is a function of the location of the sun in the sky and the clearness of the atmosphere, as well as the nature and orientation of the glazing, and presence and type of blinds and other solar control devices.

– Outdoor weather conditions

Outdoor weather conditions such as wet and dry bulb temperatures, wind speed, humidity, and solar radiation determine the magnitude of heat losses and gains.

– Occupant behaviour

Occupant behaviour, such as temperature setting, opening of windows, orientation of blinds and curtains, influence the amount of energy used for SH.

The input units of the SH network dataset were developed taking into consideration the above factors and using the information available from the 1993 SHEU database.

Input Units for Building Envelope Thermal Characteristics

The following information on building characteristics is available in the 1993 SHEU database:

- dwelling type (single detached, double, row or terrace, duplex),
- number of storeys,
- exterior wall material types (aluminium/steel siding, brick, stucco, vinyl siding, stone/concrete, wood, log, asbestos shingle),
- number of doors,
- number of triple, double, and single glazed windows,
- total heated area excluding basement and garage,
- year dwelling was built,
- basement type and area,
- attic and roof type,
- presence and year of the improvements done to wall, roof, basement wall, and basement floor insulation,
- presence of heated basement and garage.

The 1993 SHEU database does not contain sufficient information on building envelope thermal characteristics to develop the overall heat transfer coefficients for each component of the envelope. The only information available is the type of material used in the exterior walls, but information on the thickness and type of the exterior wall and insulation materials is not given. Due to lack of direct information in the 1993 SHEU database to calculate the overall heat transfer coefficient values, the vintage of the envelope components is used as proxy for envelope thermal characteristics based on the work of Farahbakhsh (1997) and Farahbakhsh et al. (1997, and 1998).[11]

As seen in Table 4.22, the year of construction of the dwellings is reported in six categories in the 1993 SHEU database. These categories are used in the input unit dataset to denote the ages of the dwellings. The year in which the insulation improvements were conducted is given in three categories as seen in Table 4.23. If a dwelling had undertaken any insulation improvement, the age

---

[11] The relationship between the vintage of the dwellings and the overall heat transfer coefficient of each component of the envelope was studied by Farahbakhsh (1997) and Farahbakhsh et al. (1997 and 1998). A database was developed using the information available from the Modified STAR HOUSING database (Scanada Consultants, 1992), the 1993/94 "200-House Audit" project database (NRCan, 1994), and the 1994 New House Survey database (NRCan, 1996), and the thermal resistances (i.e. R-values) of building envelope components (e.g. walls, roofs, doors, windows, etc.) were categorized based on the vintage of the dwellings. It was found that R-values increase as the age of the dwellings decrease.

category of that envelope component was increased as given in Table 4.24. The wall, basement wall, basement floor, and roof age categories were included as input units using this procedure. For example, if a dwelling was built before 1941 and its roof and walls were improved in 1980, age categories of the house, wall, roof, basement wall and floor would be 1, 4, 4, 1, and 1, respectively. The input units used for building envelope thermal characteristics are given in Table 4.26 along with all other input units used in the SH NN Model.

Table 4.22.   Age categories based on the year of construction of the dwellings in the 1993 SHEU database

Years	Category for the year of construction
Before 1941	1
1941 – 1960	2
1961 – 1977	3
1978 – 1982	4
1983 – 1988	5
1989 or later	6

Table 4.23.   Age categories for the year of the insulation improvements in the 1993 SHEU database

Years	Category for the year of improvement
1977 or earlier	1
1978 – 1983	2
1984 or later	3

Table 4.24.   Age categories of the envelope components with insulation improvements

Year of the insulation improvement	Updated age category
1977 or earlier	3
1978 – 1983	4
1984 or later	5

Input Units for Building Envelope Areas

The areas of the envelope components were calculated using the available information in the 1993 SHEU database and assumptions used in other similar studies (Farahbakhsh, 1997; Farahbakhsh *et al.*, 1997 and 1998). The input units used for building envelope areas are given in Table 4.26. The

69

assumptions and calculation procedures used in determining the envelope areas are given below.

Floor Area: The total heated area (excluding basement and garage) and number of storeys of the dwellings are available in the 1993 SHEU database. The floor area was calculated by dividing the total heated area by the number of storeys.

Wall Area: The wall area was calculated using Equation 4.3, assuming that the dwellings have square foot print and the height of the walls are 2.5 m. The number of walls is four for single detached, and three for single attached dwellings.

$$\text{Wall Area} = \sqrt{\text{Floor Area}} * \text{Wall Height} * \text{No. of storeys} * \text{No.of Walls} \quad (4.3)$$

Basement Floor and Wall Area: The basement type and floor area are available in the 1993 SHEU database. Assuming that basement wall height is 2.5 m for full basements, 1.8 m for partial basements, and 0.6 m for crawl spaces, the basement wall areas were calculated using Equation 4.4:

$$\text{Basement Wall Area} = \sqrt{\text{Basement Area}} * \text{Basement Wall Height} * 4 \quad (4.4)$$

Roof Area: The 1993 SHEU database contains information on the presence of an attic. If the dwelling has an attic, it was assumed that the ceiling is insulated, and the heat transfer area is taken to be equal to floor area of the dwelling. If the dwelling does not have an attic, then the roof area was calculated assuming that the dwelling has square foot print and a roof slope of 0.25. Therefore, Equation 4.5 gives the roof area:

$$\text{Roof Area} = \text{Floor Area} * 1.16 \quad (4.5)$$

Input Units for Windows and Doors

In the 1993 SHEU database, the number of single, double, triple windows and skylights, and the number of metal, wood, and patio doors are given. There is no information about the R-values, dimensions, or directions of the windows and doors. Therefore, as shown in Table 4.26, the number of doors and windows are used as input units representing the heat loss through these components.

Triple, double, and single glazed patio doors and skylights were considered as windows, and single glazed windows with storm windows were considered as double glazed windows.

Input Units for Heated Basements and Garages

The 1993 SHEU database contains qualitative information on the amount of heated basement area, which was converted to percentage values as shown in Table 4.25. Binary variables zero and one are used as input units for the heated garage ownership; the variable one is used if the household owns a heated garage, and zero is used if it does not. These input units are given in Table 4.26.

Table 4.25. Amount of basement area heated

Information in the 1993 SHEU database	Corresponding Percentage (%)
The whole basement	100
More than one half	75
About one half	50
Less than half	25

Input Unit for Dwelling Types

In the 1993 SHEU database, dwellings are classified into four categories: single detached, double, row or terrace, and duplex. In this study, double, row or terrace, and duplex dwellings are combined into one category called "single attached". Binary variables zero and one are used as input units for the dwelling type; variable one is used if the dwelling is single-detached, and zero if it is single-attached, as shown in Table 4.26.

Input Unit for Dwelling Types

In the 1993 SHEU database, dwellings are classified into four categories: single detached, double, row or terrace, and duplex. In this study, double, row or terrace, and duplex dwellings are combined into one category called "single attached". Binary variables zero and one are used as input units for the dwelling type; variable one is used if the dwelling is single-detached, and zero if it is single-attached, as shown in Table 4.26.

71

Table 4.26. SH network input units

	Input Unit	Range
Dwelling Characteristics	Dwelling type: 1 if single-detached; 0 if single-attached	0 – 1
	Number of doors	1 – 11
	Number of triple glazed windows	0 – 30
	Number of double glazed windows	0 – 48
	Number of single glazed windows	0 – 24
	Wall area [m^2]	71 - 733
	Floor area [m^2]	17 - 265
	Basement wall area [m^2]	0 - 163
	Basement floor area [m^2]	0 - 265
	Roof area [m^2]	17 - 265
	Dwelling age category	1 - 6
	Wall age category	1 - 6
	Roof age category	1 - 6
	Basement wall age category	0 - 6
	Basement floor age category	0 - 6
	Percentage of the basement heated [%]	0 – 100
	Heated garage: 1 if heated; 0 if not heated	0 – 1
SH system and equipment properties	End-use efficiency of the SH equipment [%]	65 – 100
	Presence of heat recovery ventilation system	0 – 1
	Presence of programmable thermostats	0 – 1
Indoor and outdoor temperatures	Average indoor temperature [^{o}C]	16 – 24
	Heating degree days [^{o}C-day]	2,930 – 6,541
Socio-economic characteristics of the households	Income [$1,000/yr]	10 – 85
	Dwelling ownership: 1 if owner; 0 if renter	0 – 1
	Number of children	0 - 6
	Number of adults	1 - 6
	Daytime occupancy	0 – 1
	Size of area of residence: 1 if population is less than 15,000 2 if population is between 15,000 and 100,000 3 if population is 100,000 or over	1 – 3

Input Units for SH System and Equipment Properties

The information available in the 1993 SHEU database on the SH system and equipment properties includes:

- type of SH equipment (furnace with hot air vents, boiler with hot water radiators, wood stove, electric baseboards, electric radiant heaters)
- fuel types and energy sources used for SH (natural gas, oil, electricity, wood, propane, coal)
- efficiency rating of the oil and natural gas SH equipment (standard, medium, high)
- age of the SH equipment,
- use of heat pump and back-up furnace,
- source, age, and power of heat pump system,
- fuel types and energy sources of the back-up furnaces (natural gas, oil, electricity, wood, propane, coal),
- use of programmable thermostat,
- use of heat recovery ventilation (HRV) system.

The 1993 SHEU database contains information on the efficiency ratings of the oil and natural gas fuelled SH equipment in three categories, which are standard (50-65%), medium (75-80%), and high (90% or higher) efficiency. The typical efficiency values given in HOT2000 (NRCan, 1996) for different types of oil and natural gas fuelled SH equipment are given in Table 4.27. Using the average values of the responses in the 1993 SHEU database, the HOT2000 default values from Table 4.27, and engineering judgment, the SH equipment efficiency values given in Table 4.28 were chosen to represent each efficiency rating, and energy source and fuel type. The end-use efficiency of the SH systems that use electricity were set to 100%. The input unit for the SH equipment efficiency is given in Table 4.27.

73

Table 4.27. Typical efficiency values for natural gas and oil SH equipment (NRCan, 1996)

Fuel Type	Equipment Type	Efficiency (%)
Natural Gas	1. Furnace/boiler with continuous pilot	78
	2. Furnace/boiler with spark ignition	78
	3. Furnace/boiler with spark ignition, vent damper	78
	4. Induced draft fan furnace/boiler	80
	5. Condensing furnace/boiler	94
Oil	1. Furnace/boiler	71
	2. Furnace/boiler with flue vent damper	71
	3. Furnace/boiler with flame retention head	83
	4. Mid efficiency furnace/boiler (no dilution air)	85
	5. Direct vent, non-condensing	87
	6. Condensing furnace/boiler (no chimney)	93

Table 4.28. Efficiency values used for each efficiency rating and fuel type

	Standard	Medium	High
Natural Gas	70 %	78 %	94 %
Oil	65 %	75 %	93 %

There are only a few households using heat pump systems for SH in the 1993 SHEU database. More than half of these households do not have information on the source and/or the power of their heat pump systems. Thus, the households with heat pump systems were excluded from the SH network dataset as explained in Appendix D. The programmable thermostat and HRV system ownerships are represented by binary variables zero and one as given in Table 4.26.

Input Units for Indoor and Outdoor Temperatures

The 1993 SHEU database contains information on the average indoor temperature during daytime (6 am – 6 pm), evening (6 pm – 10 pm), and overnight (10 pm – 6 am). The average indoor temperature was calculated using Equation 4.6.

74

$$AIT = \left(T_1 * \frac{12}{24}\right) + \left(T_2 * \frac{4}{24}\right) + \left(T_3 * \frac{8}{24}\right) \qquad (4.6)$$

where,

AIT: average indoor temperature [$^\circ$C]

T_1: daytime (6 am – 6 pm) temperature [$^\circ$C]

T_2: evening (6 pm – 10 pm) temperature [$^\circ$C]

T_3: overnight (10 pm – 6 am) temperature [$^\circ$C]

The outside temperature has an important effect on the SH energy consumption. Therefore, heating degree-day (HDD) is used as an input unit as shown in Table 4.26. For this purpose, 1993 HDD data for the cities in which the households in the SH network dataset are located were obtained from Environment Canada (Environment Canada, 1999).

Other Input Units

The other input units included in the dataset are the household income, dwelling ownership information, number of children and adults in the households, daytime occupancy, and size of residence area information. The 1993 SHEU database contains information if the dwellings are occupied on average weekdays. Thus, binary variables one and zero are used to indicate if the dwellings are occupied on average weekdays as given in Table 4.26 (one indicates that dwelling is occupied, zero indicates unoccupied).

Output Unit

The annual electricity, natural gas, and oil consumption for SH is the output unit for the SH network. Thus, the annual SH electricity, natural gas, and oil consumption values obtained from the energy billing data for the 1,228 households are used as the output unit of the SH network. The SH consumption data of the households covers the calendar year of 1993, and were all converted to GJ.

2.3.3. Development of the SH Network Training and Testing Datasets

To facilitate the development of the SH NN Model, the SH network dataset was divided into two subsets. One of these subsets was used for training (training set) and the other was used for testing

(testing set) of the network. The training set contains 921 households (75% of all households in the dataset) and the testing set contains 307 households (25% of all households in the dataset). The households in each sub-set were chosen randomly.

2.3.4. Selection of Activation Functions and Scaling Intervals

The training and the testing datasets of the SH network were scaled to the intervals given in Section 3.4 of Chapter 3 resulting in eleven datasets. Each of the eight configurations given in Table 3.1 was tested for each of the eleven datasets that were scaled to different intervals. Thus, a total of 88 networks were tested (= 11 scaling intervals x 8 network configurations).

A network with 28 input, two hidden, and one output units (28:2:1) trained by standard backpropagation learning algorithm with a learning rate of 0.02 was used to compare the various scaling intervals and activation functions with respect to their prediction performance.

The training of the 28:2:1 network was halted when the testing set SSE value stopped decreasing and started to increase, which is an indication of overtraining. The prediction performance and the number of cycles for each of the eight configurations that produced the best prediction performance for each of the eleven datasets are given in Table 4.29.

As seen in Table 4.29, normalization of the data produced poor predictions, which was also seen during the development of the ALC and DHW networks. The [–0.5 to 0.5], [–1.0 to 1.0], and [–0.9 to 0.9] scaling intervals also resulted in poor predictions with R^2 around 0.5. The [0.0 to 1.0], and [0.1 to 0.9] scaling intervals produced good predictions with R^2 higher than 0.85. The network with the best prediction performance (R^2 of 0.9064) used the data scaled to interval [0.1 to 0.9], the identity function for the hidden layers, and the logistic function for the output layers. Thus, in the rest of the SH network development, the identity function was used as the activation function for the hidden layers, the logistic function was used as the activation function for the output layer, and all data in the dataset were scaled to interval [0.1 to 0.9].

Table 4.29. Comparison of scaling intervals and activation functions

Scaling Interval	Applied to	Network Configuration	R^2	Cycles
0.1 to 0.9	All data	Network G: Identity + Logistic	0.9064	2260
0.1 to 0.9	Only continuous	Network G: Identity + Logistic	0.9063	1564
-0.5 to 0.5	All data	Network H: Identity + TanH	0.5062	1259
-0.5 to 0.5	Only continuous	Network H: Identity + TanH	0.5055	1282
0.0 to 1.0	All data	Network G: Identity + Logistic	0.8517	1883
-1.0 to 1.0	All data	Network E: TanH + TanH	0.5122	1087
-1.0 to 1.0	Only continuous	Network H: Identity + TanH	0.4991	195
-0.9 to 0.9	All data	Network E: TanH + TanH	0.4979	514
-0.9 to 0.9	Only continuous	Network C: Logistic + Identity	0.4941	1276
Normalized	All dataset	Network B: Logistic + TanH	0.3175	45
Normalized	Only continuous	Network H: Identity + TanH	0.3340	24

2.3.5. Development of the SH Network Architecture

There are 28 input data units and one output data unit in the SH NN Model. In order to find the number of hidden layer units resulting in the best prediction performance, networks with the number of hidden layer units ranging from one to 40 were trained with the four different learning algorithms presented in Section 2.3.1 of Chapter 2. Thus, a total of 160 networks were tested as follows: 4 learning algorithms x 40 network configurations = 160 networks. The parameters of the learning algorithms used in the analysis are given in Table 4.30.

Table 4.30. Parameters of the learning algorithms used

Learning Algorithm	Parameters*
Standard Backpropagation	$\eta : 0.0075$
Enhanced Backpropagation	$\eta : 0.0075, \mu : 0.0005, c : 0.005$
Quickprop	$\eta : 0.001, \rho : 2.0, v : 0.00001$
Resilient Propagation	$\beta : 1.1, \phi_{initial} : 0.06, \phi_{max} : 10$

*The definitions for the parameters of the learning algorithms are given in Appendix A.

Training was halted when the testing set SSE value stopped decreasing and started to increase, which is an indication of overtraining. The prediction performance in terms of SSE, R^2, RMS, and CV of the networks with the lowest testing set SSE values amongst the 40 networks for each of the four learning algorithms is presented in Table 4.31.

As seen from Table 4.31, the learning algorithms produced good predictions within the range of R^2 of 0.907 to 0.908. The network trained with the Resilient Propagation learning algorithm with two hidden layer units resulted in the lowest testing set SSE, RMS, and CV, and highest R^2, indicating that this network has the highest prediction performance amongst the networks tested with the four learning algorithms given in Table 4.30. As seen in Table 4.30, the Resilient Propagation and Quickprop learning algorithms achieved convergence at lower number of cycles than the Standard and Enhanced Backpropagation learning algorithms due to the their mechanism of updating weights and biases as mentioned in Section 2.3.1 of Chapter 2 and Appendix A.

Table 4.31. Performance of networks trained using four different learning algorithms

Network	Learning Algorithm	Number of Hidden Units	SSE	R^2	RMS	CV	Number of Cycles
28:2:1	Resilient Propagation	2	5.400	0.908	0.133	1.871	42
28:28:1	Quickprop	28	5.433	0.908	0.133	1.877	411
28:1:1	Standard Backprop.	1	5.508	0.907	0.134	1.890	3981
28:1:1	Enhanced Backprop.	1	5.508	0.907	0.134	1.890	3863

To determine the network architecture that produces the best prediction performance, different network architectures with a total of two hidden layer units in one, two or three layers were trained with the Resilient Propagation learning algorithm. The prediction performances of these networks are given in Table 4.32. None of these configurations resulted in better prediction performance than the network with two hidden units in one hidden layer. Thus, the network with one hidden layer with two units (28:2:1) trained with the Resilient Propagation learning algorithm, and using identity function for the hidden layer, the logistic function for the output layer, and dataset scaled to interval [0.1 to 0.9] was found to be the most suitable network architecture to predict the SH energy consumption in Canadian single family households. The architectural

78

configuration of the SH NN Model (28:2:1), as well as the values of the weights and biases, are given in Appendix E.

Table 4.32. Performance of networks with different architectures

Network	Number of Hidden Layer Units			SSE	R^2	RMS	CV	Number of Cycles
	Layer 1	Layer 2	Layer 3					
28:2:1	2			5.400	0.908	0.133	1.871	42
28:2:2:1	2	2		5.482	0.907	0.134	1.885	29
28:2:2:2:1	2	2	2	5.481	0.907	0.134	1.885	374
28:1:1:1	1	1		5.492	0.907	0.134	1.887	441

3. Development of the CDA Model

The CDA Model consists of three components:
- Electricity model (EM),
- Natural gas model (NGM),
- Oil model (OM).

The processes used in the development of the electricity, natural gas, and oil models are presented in this section.

3.1. Electricity Model

3.1.1. Development of the Electricity Model Dataset

The dataset used in the development of the CDA EM is a subset of the 1993 SHEU database. There are 2,050 households in the 1993 SHEU database with electricity billing data. The electricity bills of these households account for the energy consumption for major, minor, and miscellaneous appliances, lighting, space and DHW heating, and space cooling. The electricity billing data for these 2,050 households were used in the development of the CDA EM. Details of the analyses of the electricity billing data are given in Appendix B.

79

3.1.2. Development of the Electricity Model UEC Equations

The UEC equations for each end-use was developed using the information available from the 1993 SHEU database and 1993 weather and ground temperature data obtained from Environment Canada (Environment Canada, 1999). The CDA EM was developed by combining the UEC equations of the end-uses as given in Equation 4.7.

$$HEC_i = Cons \tan t + \sum_{j=1}^{N} UEC_{ij} \qquad (4.7)$$

where,

HEC_i: Electricity consumption of household i [kWh/yr]

UEC_{ij}:   End-use j unit energy consumption of household i [kWh/yr]

N: Number of electricity end-uses, *i.e.* main and supplementary space heating, DHW heating, space cooling, lighting, major and minor appliances.

The constant term in Equation 4.7 was included in the CDA EM to represent the electricity consumption by miscellaneous appliances. As seen in Equations 4.8 to 4.40, the CDA Model was developed as a linear model; however, a large variety of mathematically manipulated variables (for example by taking a power or a logarithm of one or more variables, or by cross multiplication of several variables) can be used. Use of such manipulated variables may improve the prediction performance of the CDA models.

<u>Main and Supplementary Space Heating (SH) UEC Equations</u>

The input variables of the SH UEC equations were chosen considering the structural features of the dwellings, economic and demographic characteristics of the occupants, and the weather conditions. The main and supplementary SH UEC equations used in the CDA EM are given in Equations 4.8 and 4.9, respectively, and the definitions of the commonly used variables in the equations are given in Table 4.33.

$$UEC_{SH} = SH * [a_0 + a_1 \, PROGT + a_2 \, HRV + a_3 \, AIT + a_4 \, DTYPE + a_5 \, AREA + a_6 \, AGECAT +$$

$$a_7 \, BSMNT + a_8 \, GARAGE + a_9 \, ATTIC + a_{10} \, TRIPLE + a_{11} \, DOUBLE +$$

$$a_{12} \, SINGLE + a_{13} \, DOOR + a_{14} \, HDD + a_{15} \, OWNER + a_{16} \, INCOME +$$

$$a_{17} \, CHILD + a_{18} \, ADULT + a_{19} \, DAYTIME + a_{20} \, POPUL] \qquad (4.8)$$

where,

UEC$_{SH}$:   Space heating unit energy consumption [kWh/household/yr]

SH: Dummy variable: one if the household has electricity SH equipment, zero if not.

$a_0, \ldots, a_{20}$:   Regression coefficients of each variable[12].

$$UEC_{SSH} = SSH * [a_0 + a_1 \, AIT + a_2 \, AREA + a_3 \, HDD + a_4 \, CHILD +$$
$$a_5 \, ADULT + a_6 \, DAYTIME] \tag{4.9}$$

where,

UEC$_{SSH}$:   Supplementary space heating unit energy consumption [kWh/household/yr]

SSH: Dummy variable: one if the household has electricity supplementary SH equipment, zero if not.

$a_0, \ldots, a_6$:   Regression coefficients of each variable.

To simplify the UEC equations, the variables representing the number of single, double, and triple glazed windows were combined into one variable that represents the total number of windows (WINDOW). Similarly, the variables representing the number of adults and children were combined into one variable that represents the total number of occupants (HHSIZE). These two new variables were used in place of the corresponding variables in the end-use UEC equations.

DHW Heating UEC Equation

The input variables of the DHW heating UEC equation were chosen based on the available information on DHW heating system and equipment properties, DHW consumption patterns, economic and demographic characteristics of the occupants, and the weather conditions. The DHW heating UEC equation used in the CDA EM is given in Equation 4.10, and the definitions of the variables used in the equation are given in Table 4.33.

---

[12] The a_x is used as a generic symbol to denote all regression coefficients in this work. Thus, a_1 in one given equation has a different value than another a_1 in another equation.

Table 4.33. Definitions of commonly used variables in UEC equations

Variable	Definition	Range
PROGT	Dummy variable: one if the household has programmable thermostat, zero if not.	0 - 1
HRV	Dummy variable: one if the household has a heat recovery ventilation system, zero if not.	0 - 1
EFF	Efficiency of the natural gas or oil furnace/boiler [%]	65 - 94
SHAGE	Age of the natural gas or oil furnace/boiler [years]	1 - 25
AIT	Average indoor temperature calculated using Equation 4.6 [°C]	16 - 24
DTYPE	Dummy variable: one if the dwelling type is single detached, zero if it is single attached.	0 - 1
AREA	Heated living area [m^2]	51 - 502
AGECAT	Dwelling construction year category as given in Table 4.22.	1 - 6
BSMNT	Dummy variable: one if the household has a heated basement, zero if not.	0 - 1
GARAGE	Dummy variable: one if the household has a heated garage, zero if not.	0 - 1
ATTIC	Dummy variable: one if the household has an attic, zero if not.	0 - 1
TRIPLE	Number of triple glazed windows	0 – 30
DOUBLE	Number of double glazed windows	0 – 60
SINGLE	Number of single glazed windows	0 - 74
DOOR	Number of doors	1 - 11
HDD	Heating degree days [°C-day]	2,930 – 6,541
OWNER	Dummy variable: one if owner, zero if renter.	0 - 1
INCOME	Household income [$10,000/yr]	10 - 85
CHILD	Number of children	0 - 7
ADULT	Number of adults	1 - 8
DAYTIME	Dummy variable: one if the dwelling is occupied daytime during weekdays, zero if not.	0 - 1
POPUL	Size of area of residence: 1 if population is less than 15,000 2 if population is between 15,000 and 100,000 3 if population is 100,000 or over	1 – 3
TANK	Size of the DHW tank [L]	130 - 280
SYSAGE	Age of the DHW heating system [years]	0.5 - 18
BLANKET	Dummy variable: one if there is an add-on insulation blanket around the outside of the DHW tank, zero if not.	0 - 1
PIPEINS	Dummy variable: one if there is insulation around the DHW pipes, zero if not.	0 - 1
AERATOR	Number of aerators	0 – 7

Table 4.33. (continued) Definitions of commonly used variables in UEC equations

LOWFLOW	Number of low-flow shower heads	0 - 3
GT	Ground temperature [°C]	4 - 12
CWLOAD	Clothes washer [loads/week]	0 – 15
DWLOAD	Dish washer [loads/week]	0 – 15
CACUSE	Central A/C unit usage [hours/year]	0 – 1,125
WACUSE	Window A/C unit usage [hours/year]	0 – 1,125
VOLR1	Volume of the main refrigerator [L]	0 - 625
VOLR2	Volume of the second refrigerator [L]	0 - 625
FROSTR1	Dummy variable: one if the main refrigerator is frost-free, zero if not.	0 - 1
FROSTR2	Dummy variable: one if the second refrigerator is frost-free, zero if not.	0 - 1
HHSIZE	Number of occupants in the household	1 - 11
VOLF1	Volume of the main freezer [L]	0 – 710
VOLF2	Volume of the second freezer [L]	0 - 710
MICROW	Dummy variable: one if the household has a microwave, zero if not.	0 - 1
CDLOAD	Clothes dryer [loads/week]	0 – 15
LIGHTS	Total number of incandescent, fluorescent, and halogen lamps	7 - 132
WINDOW	Total number of single, double, and triple glazed windows	4 - 76

$$UEC_{DHW} = DHW * [a_0 + a_1 \ TANK + a_2 \ SYSAGE + a_3 \ BLANKET +$$
$$a_4 \ PIPEINS + a_5 \ LOWFLOW + a_6 \ AERATOR + a_7 \ GT +$$
$$a_8 \ CWLOAD + a_9 \ DWLOAD + a_{10} \ DTYPE +$$
$$a_{11} \ OWNER + a_{12} \ INCOME + a_{13} \ CHILD +$$
$$a_{14} \ ADULT] \qquad\qquad (4.10)$$

where,

 UEC$_{DHW}$:   DHW heating unit energy consumption [kWh/household/yr]

 DHW: Dummy variable: one if the household has electricity DHW heating equipment, zero if not.

 a_0, ..., a_{14}: Regression coefficients of each variable.

Space Cooling UEC Equations

The input variables of the central and window A/C UEC equations were chosen based on the available information on space cooling system usage, structural features of the dwellings, economic and demographic characteristics of the occupants, and the weather conditions. The central and window A/C UEC equations used in the CDA EM are given in Equations 4.11 and 4.12, respectively, and the definitions of the variables used in the equations are given in Table 4.33.

$$UEC_{CAC} = CAC * [a_0 + a_1 \; CACUSE + a_2 \; DTYPE + a_3 \; AREA + a_4 \; AGECAT +$$
$$a_5 \; ATTIC + a_6 \; TRIPLE + a_7 \; DOUBLE + a_8 \; SINGLE +$$
$$a_9 \; DOOR + a_{10} \; CDD + a_{11} \; OWNER + a_{12} \; INCOME +$$
$$a_{13} \; CHILD + a_{14} \; ADULT + a_{15} \; DAYTIME] \qquad (4.11)$$

where,

 UEC$_{CAC}$:   Central A/C unit energy consumption [kWh/household/yr]

 CAC   Dummy variable: one if the household has central A/C unit, zero if not.

 a_0, ..., a_{15}: Regression coefficients of each variable.

$$UEC_{WAC} = WAC * [a_0 + a_1 \; WACUSE + a_2 \; AREA + a_3 \; AGECAT +$$
$$a_4 \; CDD + a_5 \; INCOME + a_6 \; CHILD + a_7 \; ADULT +$$
$$a_8 \; DAYTIME] \qquad (4.12)$$

where,

 UEC$_{WAC}$:   Window A/C unit energy consumption [kWh/household/yr]

 WAC   Dummy variable: one if the household has window A/C unit, zero if not.

 a_0, ..., a_8: Regression coefficients of each variable.

Major Appliances UEC Equations

The major appliances include the main and secondary refrigerators, main and secondary freezers, electric ranges, dishwashers, clothes washers, and electric clothes dryers. The input variables of the major appliances UEC equations were chosen based on the available information on appliance properties, and economic and demographic characteristics of the occupants. The major appliances UEC equations used in the CDA EM are given in Equations 4.13-4.20, and the definitions of the variables used in the equations are given in Table 4.33.

$$UEC_{REF1} = REF1 * [a_0 + a_1 \ VOLR1 + a_2 \ FROSTR1 + a_3 \ INCOME + a_4 \ HHSIZE] \tag{4.13}$$

$$UEC_{REF2} = REF2 * [a_0 + a_1 \ VOLR2 + a_2 \ FROSTR2 + a_3 \ INCOME + a_4 \ HHSIZE] \tag{4.14}$$

$$UEC_{FREZ1} = FREZ1 * [a_0 + a_1 \ VOLF1 + a_2 \ INCOME + a_3 \ HHSIZE] \tag{4.15}$$

$$UEC_{FREZ2} = FREZ2 * [a_0 + a_1 \ VOLF2 + a_2 \ INCOME + a_3 \ HHSIZE] \tag{4.16}$$

$$UEC_{COOK} = COOK * [a_0 + a_1 \ HHSIZE + a_2 \ MICROW] \tag{4.17}$$

$$UEC_{DISH} = DISH * [a_0 + a_1 \ DWLOAD] \tag{4.18}$$

$$UEC_{CLOTH} = CLOTH * [a_0 + a_1 \ CWLOAD] \tag{4.19}$$

$$UEC_{DRYER} = DRYER * [a_0 + a_1 \ CDLOAD] \tag{4.20}$$

where,

UEC_{REF1}:	Main refrigerator unit energy consumption [kWh/household/yr]
UEC_{REF2}:	Secondary refrigerator unit energy consumption, [kWh/household/yr]
UEC_{FREZ1}:	Main freezer unit energy consumption [kWh/household/yr]
UEC_{FREZ2}:	Secondary freezer unit energy consumption [kWh/household/yr]
UEC_{COOK}:	Electric range unit energy consumption [kWh/household/yr]
UEC_{DISH}:	Dishwasher unit energy consumption [kWh/household/yr]
UEC_{CLOTH}:	Clothes washer unit energy consumption [kWh/household/yr]
UEC_{DRYER}:	Electric clothes dryer unit energy consumption [kWh/household/yr]
REF1:	Dummy variable: one if the household has a refrigerator, zero if not.
REF2:	Dummy variable: one if the household has a secondary refrigerator, zero if not.
FREZ1:	Dummy variable: one if the household has a freezer, zero if not.
FREZ2:	Dummy variable: one if the household has a secondary freezer, zero if not.
COOK:	Dummy variable: one if the household has an electric range, zero if not.

DISH: Dummy variable: one if the household has a dishwasher, zero if not.

CLOTH: Dummy variable: one if the household has a clothes washer, zero if not.

DRYER: Dummy variable: one if the household has an electric clothes dryer, zero if not.

$a_0, ..., a_4$:   Regression coefficients of each variable.

Minor Appliances UEC Equations

The minor appliances include microwaves, color TVs, black and white TVs, VCRs, furnace fans and boiler pumps. The input variables of the minor appliances UEC equations were chosen based on the available information on the demographic characteristics of the occupants and weather conditions. The minor appliances UEC equations used in the CDA EM are given in Equations 4.21-4.26, and the definitions of the variables used in the equations are given in Table 4.33.

$$UEC_{FF} = FF * [a_0 + a_1 \, AREA + a_2 \, HDD] \tag{4.21}$$

$$UEC_{BP} = BP * [a_0 + a_1 \, AREA + a_2 \, HDD] \tag{4.22}$$

$$UEC_{MICROW} = MICROW * [a_0 + a_1 \, HHSIZE] \tag{4.23}$$

$$UEC_{CTV} = CTV * [a_0 + a_1 \, HHSIZE] \tag{4.24}$$

$$UEC_{BWTV} = BWTV * [a_0 + a_1 \, HHSIZE] \tag{4.25}$$

$$UEC_{VCR} = VCR * [a_0 + a_1 \, HHSIZE] \tag{4.26}$$

where,

UEC_{FF}:   Furnace fan unit energy consumption [kWh/household/yr]

UEC_{BP}:   Boiler pump unit energy consumption [kWh/household/yr]

UEC_{MICROW}:   Microwave unit energy consumption [kWh/household/yr]

UEC_{CTV}:   Color TV unit energy consumption [kWh/household/yr]

UEC_{BWTV}:   Black and white TV unit energy consumption [kWh/household/yr]

UEC_{VCR}:   VCR unit energy consumption [kWh/household/yr]

FF: Dummy variable: one if the household has a furnace, zero if not.

BP: Dummy variable: one if the household has a boiler, zero if not.

MICROW: Dummy variable: one if the household has a microwave, zero if not.

CTV: Number of colour TVs owned by the household

BWTV: Number of black and white TVs owned by the household

VCR: Number of VCRs owned by the household

$a_0, ..., a_2$:   Regression coefficients of each variable.

Lighting UEC Equation

The input variables of the lighting UEC equation are the number of halogen, incandescent, and fluorescent lamps. The lighting UEC equation used in the CDA EM is given in Equation 4.27.

$$UEC_{LIGHT} = a_1 \ HALO + a_2 \ INCA + a_3 \ FLOU \qquad (4.27)$$

where,

UEC_{LIGHT}:	Lighting unit energy consumption [kWh/household/yr]
HALO:	Number of halogen lamps owned by the household
INCA:	Number of incandescent lamps owned by the household
FLOU:	Number of fluorescent lamps owned by the household
$a_1, \ldots, a_3$:	Regression coefficients of each variable.

To simplify the UEC equations, the variables representing the number of halogen, incandescent, and fluorescent lamps were combined into one variable that represents the total number of lamps (LIGHTS). This new variable was used to replace the corresponding variables in Equation 4.27.

3.1.3. Estimation of the Electricity Model Equation

The end-use UEC equations given in Equations 4.8 to 4.27 were combined to develop the CDA EM. The model equation resulting from the integration of all individual end-use UEC equations is given in Figure F.1 of Appendix F. SYSTAT statistical software (SYSTAT, 1998) was used to regress the CDA EM equation, and the detailed output of the regression analysis is given in Figure F.2 of Appendix F. The model equation was reduced by removing the non-significant variables at 10% level, outlier and influential data points, and variables that increase multicollinearity. The resulting model equation is as follows:

$$HEC = 2128.65 + SH * [1.60 * HDD - 2516.01 * HRV + 1891.75 * DTYPE +$$
$$12.67 * AREA - 785.40 * AGECAT +$$
$$78.14 * INCOME] +$$
$$SSH * [8.13 * AREA + 569.33 * CHILD] +$$
$$DHW * [16.86 * TANK - 691.34 * LOWFLOW +$$
$$215.04 * DWLOAD + 752.83 * ADULT] +$$
$$CAC * [1.27 * CACUSE] +$$
$$REF1 * [1030.23 * FROSTR1] +$$
$$REF2 * [1636.48 * FROSTR2 + 28.24 * INCOME] +$$
$$FREZ1 * [1.50 * VOLF1] + COOK * [421.25 * HHSIZE] +$$
$$DRYER * [303.61 * CDLOAD] + 50.18 * LIGHTS \qquad (4.28)$$

As shown in Figure F.2 of Appendix F, the multiple coefficient of determination of the CDA EM is 0.66, which indicates that 66% of variation in the estimated household electricity consumption can be accounted for by the prediction based on the variables of the Equation 4.28.

As stated in Section 4.4 of Chapter 3, one way to measure multicollinearity is determining the condition indices (Weisberg, 1985). As shown in Figure F.3 of Appendix F, the condition index of the model is 15.865 suggesting a possible multicollinearity problem (SYTAT, 1998). The condition index of the model can be reduced by removing the variables that are highly correlated with others. However, the model has already been reduced to a minimum number of variables required to distinguish the electricity consumption into end-uses. Further reduction in the number of variables could restrict the capability of the model to estimate the consumption of major electricity end-uses.

With the multicollinearity problem, the estimation of the coefficients would still be unbiased (Johnston and DiNardo, 1997), but inferences from the standard errors are likely to be misleading. In other words, multicollinearity problem would not affect the predictions as long as estimates of the average household electricity consumption would be in the range of the original data. The derivations to prove that the estimations of the coefficients are still unbiased under multicollinearity are given in Appendix G.

As seen in Figure 4.1, the residuals (*i.e.* the errors) form an approximate straight line in the normal probability. This confirms that the errors are distributed normally.

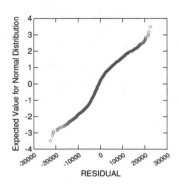

Figure 4.1. Normal probability plot of the residuals of the CDA EM

In order to check the assumption that the variability of the residuals is the same at all levels, the residuals were plotted against the estimated values as given in Figure 4.2. The plot shows that the errors might not have constant variance since the errors were not arranged in a horizontal band within two or three units around the zero line of the plot; instead a right opening megaphone structure was formed.

The White test (Johnston and DiNardo, 1997) was used to check the non-constant variance assumption of this model. The result of the test shows that the constant variance assumption was violated. However, the power of the White test diminishes as the number of variables of the model increases (Johnston and DiNardo, 1997). The CDA EM has 20 variables, thus the result of the test is not conclusive. In addition, the estimation of the coefficients would still be unbiased with the non-constant variance situation (Johnston and DiNardo, 1997).

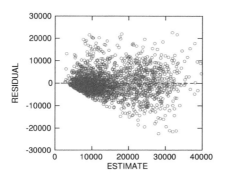

Plot of Residuals against Predicted Values

Figure 4.2. Plot of residuals of the CDA EM against estimated values

3.2. Natural Gas Model

3.2.1. Development of the Natural Gas Model Dataset

The dataset used in the development of the CDA NGM is a subset of the 1993 SHEU database. There are 1,012 households in the 1993 SHEU database with natural gas billing data, and four of these households have no natural gas end-uses reported in the 1993 SHEU database. The natural gas bills of the remaining 1,008 households account for the energy consumption for space and DHW heating, cooking, clothes drying, pool heating, and fireplaces. The natural gas billing data for these 1,008 households were used in the development of the CDA NGM. Details of the analyses of the natural gas billing data are given in Appendix B.

3.2.2. Development of the Natural Gas Model UEC Equations

The UEC equation for each end-use was developed using the information available from the 1993 SHEU database and 1993 weather and ground temperature data obtained from Environment Canada (Environment Canada, 1999). The CDA NGM was developed by combining the UEC equations of

90

the end-uses as given in Equation 4.29. Since all natural gas end-uses were reported in the 1993 SHEU database, the constant term was not included in the model.

$$HEC_i = \sum_{j=1}^{N} UEC_{ij}$$ (4.29)

where,

HEC_i:	Natural gas consumption of household i [m^3/yr]
UEC_{ij}:	End-use j unit natural gas consumption of household i [m^3/yr]
N:	Number of natural gas end-uses, *i.e.* main and supplementary space heating, DHW heating, clothes drying, cooking, pool heating, and fireplaces.

Main and Supplementary SH UEC Equations

The input variables of the SH UEC equations were chosen considering the end-use efficiency of the SH equipment, structural features of the dwellings, economic and demographic characteristics of the occupants, and the weather conditions. The supplementary SH UEC equation includes the households with natural gas supplementary SH units and fireplaces, since there are only eight households with natural gas supplementary heating systems and 37 households with natural gas fireplaces. The main and supplementary SH UEC equations used in the CDA NGM are given in Equations 4.30 and 4.31, respectively, and the definitions of the variables used in the equations are given in Table 4.33.

$$
\begin{aligned}
UEC_{SH} = SH * [& a_0 + a_1\, EFF + a_2\, SHAGE + a_3\, PROGT + a_4\, AIT + \\
& a_5\, DTYPE + a_6\, AREA + a_7\, AGECAT + a_8\, BSMNT + \\
& a_9\, GARAGE + a_{10}\, ATTIC + a_{11}\, TRIPLE + a_{12}\, DOUBLE + \\
& a_{13}\, SINGLE + a_{14}\, DOOR + a_{15}\, HDD + a_{16}\, OWNER + \\
& a_{17}\, INCOME + a_{18}\, CHILD + a_{19}\, ADULT + a_{20}\, DAYTIME + \\
& a_{21}\, POPUL]
\end{aligned}
$$ (4.30)

where,

UEC_{SH}:	Space heating unit energy consumption [m^3/household/yr]
SH:	Dummy variable: one if the household has natural gas SH equipment, zero if not.
$a_0, \ldots, a_{21}$:	Regression coefficients of each variable.

91

$$UEC_{SSH} = SSH * [a_0 + a_1 \, AIT + a_2 \, AREA + a_3 \, HDD + a_4 \, CHILD +$$
$$a_5 \, ADULT + a_6 \, DAYTIME] \qquad\qquad (4.31)$$

where,

UEC_{SSH}:	Supplementary space heating unit energy consumption [m^3/household/yr]
SSH:	Dummy variable: one if the household has natural gas supplementary SH equipment and/or fireplace, zero if not.
$a_0, ..., a_6$:	Regression coefficients of each variable.

DHW Heating UEC Equation

The input variables of the DHW heating UEC equation were chosen based on the available information on DHW heating system and equipment properties, DHW consumption patterns, economic and demographic characteristics of the occupants, and the weather conditions. The DHW heating UEC equation used in the CDA NGM is given in Equation 4.32, and the definitions of the variables used in the equation are given in Table 4.33.

$$UEC_{DHW} = DHW * [a_0 + a_1 \, TANK + a_2 \, SYSAGE + a_3 \, BLANKET +$$
$$a_4 \, PIPEINS + a_5 \, LOWFLOW + a_6 \, AERATOR + a_7 \, GT +$$
$$a_8 \, CWLOAD + a_9 \, DWLOAD + a_{10} \, DTYPE +$$
$$a_{11} \, OWNER + a_{12} \, INCOME + a_{13} \, CHILD +$$
$$a_{14} \, ADULT] \qquad\qquad (4.32)$$

where,

UEC_{DHW}:	DHW heating unit energy consumption [m^3/household/yr]
DHW:	Dummy variable: one if the household has natural gas DHW heating equipment, zero if not.
$a_0, ..., a_{14}$:	Regression coefficients of each variable.

Cooking, Clothes Drying, and Pool Heating UEC Equations

The input variables of the cooking, clothes drying, and pool heating UEC equations were chosen based on the available information on appliance usage, and economic and demographic characteristics of the occupants. The UEC equations used in the CDA NGM are given in Equations

4.33-4.35, and the definitions of the variables used in the equations are given in Table 4.33.

$$UEC_{COOK} = COOK * [a_0 + a_1 \ HHSIZE + a_2 \ MICROW] \hspace{2cm} (4.33)$$

$$UEC_{DRYER} = DRYER * [a_0 + a_1 \ CDLOAD] \hspace{2.5cm} (4.34)$$

$$UEC_{POOL} = POOL * [a_0 + a_1 \ INCOME] \hspace{2.7cm} (4.35)$$

where,

UEC_{COOK}:	Natural gas range unit energy consumption [m^3/household/yr]
UEC_{DRYER}:	Natural gas clothes dryer unit energy consumption [m^3/household/yr]
UEC_{POOL}:	Natural gas pool heating unit energy consumption [m^3/household/yr]
COOK:	Dummy variable: one if the household has a natural gas range, zero if not.
DRYER:	Dummy variable: one if the household has a natural gas clothes dryer, zero if not.
POOL:	Dummy variable: one if the household has a natural gas pool heater, zero if not.
$a_0, ..., a_2$:	Regression coefficients of each variable.

3.2.3. Estimation of the Natural Gas Model Equation

The end-use UEC equations given in Equations 4.30 to 4.35 were combined to develop the CDA NGM. The model equation resulting from the integration of all individual end-use UEC equations is given in Figure F.4 of Appendix F. The SYSTAT statistical software (SYSTAT, 1998) was used to regress the CDA NGM equation, and the output of the regression analysis is given in Figure F.5 of Appendix F. The model equation was reduced by removing the non-significant variables at 10% level, outlier and influential data points, and variables that increase multicollinearity. The resulting model equation is as follows:

$$\begin{aligned}
HEC = SH * [&-207.75 * PROGT + 129.84 * DOOR + 26.30 * WINDOW + 46.03 * SHAGE + \\
&437.69 * GARAGE + 6.61 * AREA + 181.61 * ADULT] + \\
DHW * [&13.21 * SYSAGE + 70.77 * HHSIZE + 720.97 * DTYPE] + \\
COOK * [&94.04 * HHSIZE] + DRYER * [35.91 * CDLOAD] + \\
POOL * &969.56 \hspace{4cm} (4.36)
\end{aligned}$$

As shown in Figure F.5 of Appendix F, the multiple coefficient of determination of the CDA NGM is 0.92, which indicates that 92% of variation in the estimated household natural gas consumption can be accounted for by the prediction from the variables of the Equation 4.36. The condition index of the model is 10.27 indicating that there is not a significant multicollinearity problem as given in Figure F.6 of Appendix F.

The residuals (*i.e.* the errors) form an approximate straight line in the normal probability plot as seen in Figure 4.3. This confirms that the errors are distributed normally.

In order to check the assumption that the variability of the residuals is the same at all levels, the residuals were plotted against the estimated values as given in Figure 4.4. The plot shows that the errors have constant variance since the errors were arranged in a horizontal band within two or three units around the zero line of the plot. When the White test was applied to the CDA NGM, its result showed that the constant variance assumption was violated. However, the number of variables in the CDA NGM is high, therefore the power of the test tends to diminish and its result could be unreliable (Johnston and DiNardo, 1997).

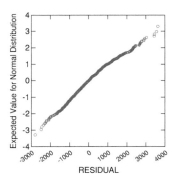

Figure 4.3. Normal probability plot of the residuals of the CDA NGM

Plot of Residuals against Predicted Values

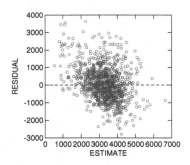

Figure 4.4. Plot of residuals of the CDA NGM against estimated values

3.3. Oil Model

3.3.1. Development of the Oil Model Dataset

The dataset used in the development of the CDA OM is a subset of the 1993 SHEU database. There are 236 households in the 1993 SHEU database with oil billing data, and five of these households have no oil end-uses reported in the 1993 SHEU database. The oil bills of the remaining 231 households account for the energy consumption for space and DHW heating, cooking, and pool heating. The oil billing data of these 231 households were used in the development of the CDA OM. Details of the analyses of the oil billing data are given in Appendix B.

3.3.2. Development of the Oil Model UEC Equations

There are only two households with oil cooking ranges or pool heaters in the database. Therefore, oil cooking and pool heating energy consumption are not addressed in the model. UEC equations for SH and DHW heating end-uses were developed using the information available from the 1993 SHEU database and 1993 weather and ground temperature data obtained from Environment Canada (Environment Canada, 1999). The CDA OM was developed by combining the UEC equations of

95

the end-uses as given in Equation 4.37. Since all oil end-uses were reported in the 1993 SHEU database, the constant term was not included in the model.

$$HEC_i = \sum_{j=1}^{N} UEC_{ij}$$
(4.37)

where,

HEC$_i$:   Oil consumption of household i [L/yr]

UEC$_{ij}$:   End-use j unit oil consumption of household i [L/yr]

N: Number of oil end-uses, *i.e.* main and supplementary space heating, DHW heating, cooking, and pool heating.

SH UEC Equation

There are only two households with oil supplementary heating units in the CDA OM dataset. Thus, the oil supplementary space heating UEC equation is excluded from the analysis. The input variables of the SH UEC equation were chosen considering the end-use efficiency of the SH equipment, structural features of the dwellings, economic and demographic characteristics of the occupants, and the weather conditions. The SH UEC equation used in the CDA OM is given in Equation 4.38, and the definitions of the variables used in the equations are given in Table 4.33.

$$
\begin{aligned}
UEC_{SH} = SH * [\, & a_0 + a_1\ EFF + a_2\ SHAGE + a_3\ PROGT + a_4\ AIT + \\
& a_5\ DTYPE + a_6\ AREA + a_7\ AGECAT + a_8\ BSMNT + \\
& a_9\ GARAGE + a_{10}\ ATTIC + a_{11}\ TRIPLE + a_{12}\ DOUBLE + \\
& a_{13}\ SINGLE + a_{14}\ DOOR + a_{15}\ HDD + a_{16}\ OWNER + \\
& a_{17}\ INCOME + a_{18}\ CHILD + a_{19}\ ADULT + a_{20}\ DAYTIME + \\
& a_{21}\ POPUL\,]
\end{aligned}
$$
(4.38)

where,

UEC$_{SH}$:   Space heating unit energy consumption [L/household/yr]

SH: Dummy variable: one if the household has oil SH equipment, zero if not.

$a_0, ..., a_{21}$:   Regression coefficients of each variable.

96

DHW Heating UEC Equation

The input variables of the DHW heating UEC equation were chosen based on the available information on DHW heating system and equipment properties, DHW consumption patterns, economic and demographic characteristics of the occupants, and the weather conditions. The DHW heating UEC equation used in the CDA OM is given in Equation 4.39, and the definitions of the variables used in the equation are given in Table 4.33.

$$UEC_{DHW} = DHW * [a_0 + a_1 \, TANK + a_2 \, SYSAGE + a_3 \, BLANKET +$$
$$a_4 \, PIPEINS + a_5 \, LOWFLOW + a_6 \, AERATOR + a_7 \, GT +$$
$$a_8 \, CWLOAD + a_9 \, DWLOAD + a_{10} \, DTYPE +$$
$$a_{11} \, OWNER + a_{12} \, INCOME + a_{13} \, CHILD +$$
$$a_{14} \, ADULT] \qquad\qquad (4.39)$$

where,

UEC_{DHW}:   DHW heating unit energy consumption [L/household/yr]

DHW: Dummy variable: one if the household has oil DHW heating equipment, zero if not.

$a_0, ..., a_{14}$:   Regression coefficients of each variable.

3.3.3. Estimation of the Oil Model Equation

The SH and DHW heating UEC equations given in Equations 4.38 and 4.39 were combined to develop the CDA OM. The model equation resulting from the integration of all individual end-use UEC equations is given in Figure F.7 of Appendix F. The SYSTAT statistical software (SYSTAT, 1998) was used to regress the CDA OM equation, and the oooutput of the regression analysis is given in Figure F.8 of Appendix F. The model equation was reduced by removing the non-significant variables at 10% level, outlier and influential data points, and variables that increase multicollinearity. The resulting model equation is as follows:

$$HEC = SH * [40.05 * SHAGE + 4.95 * AREA + 41.68 * WINDOW +$$
$$471.21 * DTYPE] +$$
$$DHW * [5.28 * TANK + 50.91 * DWLOAD] \qquad\qquad (4.40)$$

97

As shown in Figure F.8 of Appendix F, the multiple coefficient of determination of the CDA OM is 0.87, which indicates that 87% of variation in the estimated household oil consumption can be accounted for by the prediction from the variables of the Equation 4.40. The condition index of the model is 5.95 suggesting that there is no significant multicollinearity problem as given in Figure F.9 of Appendix F.

The residuals (*i.e.* the errors) form an approximate straight line in the normal probability plot as seen in Figure 4.5. This confirms that the errors are distributed normally.

In order to check the assumption that the variability of the residuals is the same at all levels, the residuals were plotted against the estimated values as given in Figure 4.6. The plot shows that the errors have constant variance since the errors were arranged in a horizontal band within two or three units around the zero line of the plot. When the White test was applied to the CDA OM, its result showed that the constant variance assumption was not violated. Since the number of variables in the CDA OM is not high, the result of the test is reliable.

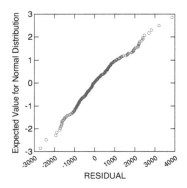

Figure 4.5. Normal probability plot of the residuals of the CDA OM

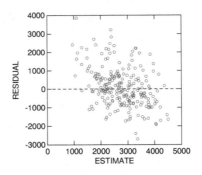

Plot of Residuals against Predicted Values

Figure 4.6. Plot of residuals of the CDA OM against estimated values

4. Closing Remarks

NN and CDA Models were developed to estimate the ALC, DHW, and SH energy consumption in single family residences in Canada. The NN Model is the first of its kind. Before this work, NNs were not used to model residential energy consumption at the national or regional level. Although the CDA modeling approach has been used to model residential energy consumption at the regional level, it was not used at the national level prior to this work. Thus, the CDA Model also represents an original addition to the state-of-the-art in energy modeling.

The ALC NN Model was developed using the available information in the subset of 988 households with electricity billing data in the 1993 SHEU database. The households in the ALC dataset do not have electrical DHW or space heating equipment, thus their electricity consumption billing data represent appliance, lighting, and space cooling electricity consumption. The model has 55 input units and three hidden layers each having nine units. The Quickprop learning algorithm was used to train the network with the training dataset of 741 households. The logistic function was used as the activation function for the hidden layers, the identity function was used as the activation function for the output layer, and all data in the dataset were scaled to interval [–0.5 to 0.5]. The ALC NN model achieved a high prediction performance ($R^2 = 0.909$) for a subset of households with electricity bills.

The DHW dataset was developed using the available information in the 1993 SHEU database of the 563 households with natural gas and electricity billing data that represent DHW heating energy consumption. For the DHW NN Model, the logistic function was used as the activation function for the hidden and output layers, and only continuous data in the datasets were scaled to interval [0.1 to 0.9]. The model has 18 input units and one hidden layer with 29 hidden units. The Resilient Propagation learning algorithm was used to train the network with the training dataset of 422 households. The DHW NN model achieved a high prediction performance ($R^2 = 0.871$) for a subset of households with natural gas and electricity bills.

The SH dataset used to develop the SH NN Model contains available information in the 1993 SHEU database of the 1,228 households with electricity, natural gas, and oil billing data. The NN Model achieved a high prediction performance ($R^2 = 0.908$) for a subset of households with electricity, natural gas, and oil bills for SH energy consumption. The model has 28 input units and one hidden layer with two units. The Resilient Propagation learning algorithm was used to train the network with the training dataset of 921 households. The identity function was used as the activation function for the hidden layer, the logistic function was used as the activation function for the output layer, and all data in the dataset were scaled to interval [0.1 to 0.9].

The DHW heating electricity consumption of the households in the DHW NN dataset were calculated by deducting the ALC electricity consumption estimated using the ALC NN Model from the total annual electricity billing data. Also, for electrically heated households, the annual SH electricity consumption was calculated by deducting the ALC and DHW heating electricity consumption estimated using the ALC and DHW NN Models from the total annual electricity billing data. For natural gas heated households, the annual SH natural gas consumption was calculated by deducting the DHW heating natural gas consumption estimated by the DHW NN Model from the total annual energy billing data. This introduces error in the DHW heating and SH energy consumption data used in developing the DHW and SH NN Models; however, this approach was used since disaggregated DHW heating and SH energy consumption data do no exist.

The CDA EM was developed with a constant term and a total of 20 variables from 12 electricity end-use UEC equations. The model achieved a multiple coefficient of determination of 0.66. The condition index of the model suggested a possible multicollinearity problem, however this problem would not effect the unbiased estimation of the coefficients. Despite the multicollinearity problem, the number of variables of the model was not reduced in order to retain the capability of the model to estimate the energy consumption for each end-use.

The graphical analysis of the residuals showed that the errors are normally distributed. However, the graphical analysis of the residuals and the White test result for the constant variance assumption showed that the residuals do not have constant variance. On the other hand, the estimation of the coefficients would still be unbiased with the non-constant variance situation (Johnston and DiNardo, 1997).

The CDA NGM was developed with a total of 13 variables from five natural gas end-use UEC equations. The CDA OM has six variables from two oil end-use UEC equations. The CDA natural gas and oil models achieved high values for multiple coefficient of determination, 0.97 and 0.82, respectively. Neither model faced the multicollinearity problem.

The CDA Model developed in this work is linear; however, a large variety of mathematically manipulated variables (for example by taking a power or a logarithm of one or more variables, or by cross multiplication of several variables) can be used. Use of such manipulated variables may improve the prediction performance of the CDA models.

It can thus be stated that the NN and CDA methods were shown to be suitable for the modeling of residential energy consumption at the national and regional levels. The models that were developed in this work demonstrated to have good prediction performance.

The comparisons of the prediction performance of the NN, CDA, and Engineering Models based on the metered energy consumption data and estimates are presented in the next chapter.

Chapter 5

Comparison of the Models

1. Overview

In this chapter, a comparative evaluation of the NN, CDA, and Engineering Models is conducted by comparing their energy consumption estimates with actual energy consumption data, as well as by comparing their estimates with each other. Comparisons are carried out for the following end-uses:

- Appliance, lighting, and space cooling (ALC) energy consumption,
- Domestic hot water (DHW) heating energy consumption,
- Space heating (SH) energy consumption.

The Engineering Model developed by Farahbakhsh (1997) and Farahbakhsh *et al.* (1997, 1998) is used in the comparisons. In addition to the comparison of the models with respect to the accuracy of their predictions, a qualitative comparison is also carried out to show the relative strengths and weaknesses of each model, and its usefulness.

The chapter continues with a review and analysis of the household energy consumption estimates of the NN, CDA, and Engineering Models based on SH energy source and fuel type, dwelling type and age, and provincial distribution. In the last section of the chapter, the effects of some socio-economic factors on the mentioned end-uses are examined.

2. ALC Energy Consumption

2.1. Households with Energy Billing Data

The prediction performance of the NN, CDA, and Engineering Models was assessed by comparing the estimates of the models with actual energy consumption data from the 247 households in the ALC NN testing dataset. The results are presented in Table 5.1. The CDA and the Engineering Models have lower R^2 and higher CV values than the NN Model, which shows that the CDA and the Engineering Models have a lower prediction performance than the NN Model.

Table 5.1. Prediction performance of the NN, CDA, and Engineering Models- ALC

Model	R^2	CV
Engineering	0.780	3.463
CDA	0.795	3.343
NN	0.909	2.094

The estimates of the Engineering, CDA, and NN Models are plotted along with the actual energy consumption data for the 247 households in the ALC NN testing dataset in Figures 5.1 to 5.3, respectively. The NN Model, as well as the CDA and Engineering Models, were not able to predict the energy consumption of some households with high energy billing data as seen in Figure 5.1 to 5.3. When the input units of these households were examined, it was found that these households could not be expected to have such high ALC electricity consumption values. The number, size, and usage of appliances of these households are very close to the corresponding average values in the testing dataset. This shows that there are other factors affecting the electricity consumption than those reported in the 1993 SHEU database and represented by the input units in the model (such as workshops in garages) that would increase ALC electricity consumption of the households.

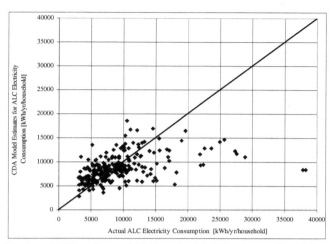

Figure 5.1. Actual billing data and Engineering Model estimates for ALC energy consumption of the households in the ALC NN testing dataset

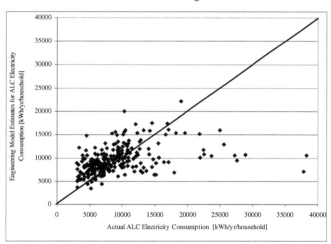

Figure 5.2. Actual billing data and CDA Model estimates for ALC energy consumption of the households in the ALC NN testing dataset

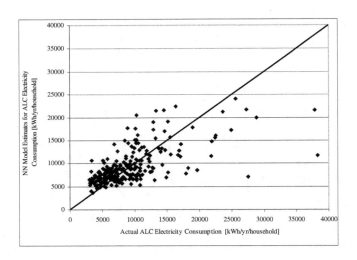

Figure 5.3. Actual billing data and NN Model estimates for ALC energy consumption of the households in the ALC NN testing dataset

2.2. Households without Energy Billing Data

2.2.1. Average Household ALC Electricity Consumption

The 1993 SHEU database has data on 8,767 households, and data on 988 of these were used to develop the ALC NN Model. These 988 households were not included in the dataset used in the predicting the various components of ALC electricity consumption. The "ALC prediction dataset" therefore includes the remaining 7,779 households. The NN and the CDA Models are used here to predict the ALC electricity consumption of these households. Since the 1993 SHEU database is representative of the Canadian housing stock, it is possible to extrapolate the predicted electricity consumption of the households in the 1993 SHEU database to the entire Canadian housing stock using weighting factors (Statistics Canada, 1993).

The weighted average ALC consumption estimated by the CDA, Engineering, and NN Models for the 7,779 households in the ALC prediction dataset, and the average percent deviations between the NN Model and Engineering Model, and between the NN Model and CDA Model calculated using Equations 5.1a and 5.1b, respectively, are given in Table 5.2. The ALC electricity

105

consumption estimated by the Engineering Model is about 2.5% higher than the NN Model estimate, whereas the CDA Model estimate is 4.6% lower.

$$\text{Deviation}(\%) = 100 * \left(\frac{\text{Enginnering Model Estimate - NN Model Estimate}}{\text{NN Model Estimate}} \right) \qquad (5.1a)$$

$$\text{Deviation}(\%) = 100 * \left(\frac{\text{CDA Model Estimate - NN Model Estimate}}{\text{NN Model Estimate}} \right) \qquad (5.1b)$$

Table 5.2. Weighted average ALC electricity consumption estimated by the Engineering, CDA, and NN Models for the 7,779 households and average percent deviations

Model	Weighted Average ALC Electricity Consumption [kWh/yr/household]	Average Deviation [%]
NN	8,791	-
CDA	8,391	-4.6
Engineering	9,012	2.5

The appliance and lighting energy consumption estimate used by NRCan's *EnerGuide for Houses* program is 24 kWh/day/household, which is equivalent to 8,760 kWh/yr/household (NRCan, 2002). This estimate does not include the household space cooling energy consumption. However, the space cooling energy consumption accounted for about 0.5% of the total national residential energy consumption in 1993 (OEE, 2002). Therefore, the addition of space cooling energy consumption to the NRCan's average household appliance and lighting energy consumption estimate of 8,760 kWh/yr/household would make an insignificant increase. As seen in Table 5.2, the estimates by all three models are close to the estimate by NRCan.

2.2.2. Scenarios to Predict Electricity Consumption of Some Appliances

The average electricity consumption of individual appliances in the ALC input unit dataset was estimated using the ALC NN Model. It was found that the NN model fails to accurately estimate the electricity consumption of appliances with high saturation values (*e.g.* main refrigerator). However, the model was able to reasonably predict the average electricity consumption of several major appliances as discussed below.

Central A/C Unit Electricity Consumption

There are 720 households with central A/C units in the ALC prediction dataset of 7,779 households. In order to estimate the central A/C electricity consumption, a dataset was developed by changing the central A/C input units from one to zero in the input files of the 720 households, and the ALC NN Model was applied to these files to estimate the household electricity consumption. Thus, as shown in Table 5.3, the difference between the predicted electricity consumption of the 720 households with and without the central A/C units provides the estimated electricity consumption by the central A/C units. As seen in Table 5.3, the average electricity consumption estimated by the ALC NN Model is 832 kWh/yr/household.

Table 5.3. Average ALC electricity consumption of 720 households in set of 7,779 households with and without central A/C units

	Average Electricity Consumption [kWh/yr/household]
720 Households with central A/C units	9,941
720 Households without central A/C units	9,109
Electricity consumption difference $\approx$ Central A/C consumption	832

As seen in Equation 4.28, the central A/C usage is one of the variables of the CDA EM. Thus, it was possible to estimate the average central A/C electricity consumption of the 720 households in the ALC prediction dataset using the CDA EM. The CDA EM estimate of 569 kWh/household/year is 32 % lower than estimate of the NN Model.

The average central A/C consumption for these 720 households was estimated by the Engineering Model to be 1,624 kWh/yr/household. However, the Engineering Model does not consider the central A/C usage characteristics (*i.e.* whether the central A/C unit is used continuously or intermittently). Based on the Engineering Model estimates, the central A/C consumption of the households in the 1993 SHEU database was estimated statistically by Aydinalp *et al.* (1998) to be 865 kWh/yr/household. This estimate took into consideration the central A/C usage characteristics of the households. The 865 kWh/yr/household estimate is slightly higher than the 832 kWh/yr/household estimate of the NN Model.

The CDA model developed by Lafrange and Perron (1994) using Hydro Quebec data estimated the central A/C consumption to be 1,662 kWh/yr/household without considering the usage characteristics of the central A/C unit. Another CDA model developed by Kellas (1993) using Manitoba Hydro data estimated the central A/C consumption to be 1,360 kWh/yr/household.

The central A/C unit electricity consumption estimates of the CDA models of Lafrange and Perron (1994) and Kellas (1993), and the Engineering Model are about 60 to 100% higher than that of the NN Model. This is largely owing to the fact that the central A/C electricity consumption was estimated by the CDA models and the Engineering Model taking into account only the central A/C unit ownership, and assuming a constant number of hours of central A/C usage. The estimates of the statistical model by Aydinalp et al. (1998), and the NN and CDA EM Models were developed considering the number of hours each household used the central A/C unit during the summer. Thus, the estimates obtained using the NN Model, as well as the CDA EM and the statistical model are more reasonable than those obtained using the CDA models by Lafrange and Perron (1994) and Kellas (1993), and the Engineering Model.

Second Refrigerator Electricity Consumption

There are 1,444 households with second refrigerators in the ALC prediction dataset of 7,779 households. In order to estimate the second refrigerator electricity consumption, a dataset was developed by changing the second refrigerator input units from one to zero in the input files of the 1,444 households, and the ALC NN Model was applied to these files to estimate the household electricity consumption. Thus, as shown in Table 5.4, the difference between the predicted electricity consumption of the 1,444 households with and without the second refrigerator provides the estimated electricity consumption by the second refrigerator. As seen in Table 5.4, the average electricity consumption estimated by the ALC NN is 755 kWh/yr/household.

Table 5.4. Average ALC electricity consumption of 1,444 households in the set of 7,779 households with and without second refrigerator

	Average Electricity Consumption [kWh/yr/household]
1,444 Households with second refrigerator	9,969
1,444 Households without second refrigerator	9,214
Electricity consumption difference ≈ Second refrigerator consumption	755

As seen in Equation 4.28, there are two variables in the CDA EM that are from the second refrigerator UEC equation (Equation 4.14). Thus, it was possible to estimate the average second refrigerator electricity consumption of the 1,444 households in the ALC prediction dataset using the CDA EM. The CDA EM estimate of 1,981 kWh/household/year is about 2.5 times higher than the estimate of NN Model. On the other hand, the CDA model developed by Kellas (1993) using Manitoba Hydro data estimated second refrigerator consumption to be 815 kWh/yr/household.

The size of the refrigerator is one of the factors affecting its electricity consumption. The average sizes for the main and second refrigerators in the SHEU 1993 database are 460 L and 375 L, respectively. Since the second refrigerators are smaller than the main refrigerators, their electricity consumption is expected to be lower than that of the main ones. Given that an average main refrigerator consumes around 1,330 kWh/yr/household (Fung et al., 1997; Ugursal and Fung, 1994), electricity consumption by the second refrigerator should be lower than 1,330 kWh/yr/household. This shows that the second refrigerator electricity consumption estimate of the CDA EM is high, while estimates of the NN Model and the CDA model developed by Kellas (1993) are reasonable.

3. DHW Heating Energy Consumption

3.1. Households with Energy Billing Data

The prediction performance of the NN, CDA, and Engineering Models was assessed by comparing the estimates of the models with actual energy consumption data from the 141 households in the DHW NN testing dataset. The results are presented in Table 5.5. The CDA and the Engineering Models have lower R^2 and higher CV values than the NN Model, which shows that the CDA and the Engineering Models have a lower prediction performance than the NN Model.

Table 5.5. Prediction performance of the NN, CDA, and Engineering Models- DHW

Model	R^2	CV
Engineering	0.828	3.898
CDA	0.814	4.052
NN	0.871	3.337

The estimates of the Engineering, CDA, and NN Models are plotted along with the actual energy consumption data for the 141 households in the DHW NN testing dataset in Figures 5.4 to 5.6, respectively. The DHW NN Model, as well as the Engineering and CDA Models, failed to accurately predict the DHW heating energy consumption of most of the households with consumption values lower than 15 GJ/yr as seen in Figures 5.4 to 5.6. When the input data for these households were analyzed, it was found that most of these households use electricity for DHW heating. As pointed out in Section 2.2.1 of Chapter 4, the amount of electricity consumed in these households for DHW heating was calculated by subtracting the ALC electricity consumption predicted by the ALC NN Model from the total billed electricity consumption. Therefore, the annual DHW heating electricity consumption of these households contain a cumulative error from the ALC and the DHW NN Models.

Out of 563 households in the DHW dataset, 388 have electricity bills, while the remaining 175 have natural gas bills. It was not possible to exclude the 391 households with electricity bills from the DHW dataset and develop a representative model with data from only 175 households with natural gas bills.

When the input units of these households were examined, it was found that based on the values of the input units these households could not be expected to have such low DHW heating energy consumption values. The age and size of the DHW heating systems, the annual ground water temperature, the number of clothes and dishwasher weekly loads, and the number of occupants of these households are either higher or very close to the corresponding average values in the testing dataset. This shows that there are other factors affecting the DHW heating energy consumption than those reported in the 1993 SHEU database and represented by the input units in the model that would decrease DHW load of the households (such as taking morning showers in the gym, or some people not taking as many showers).

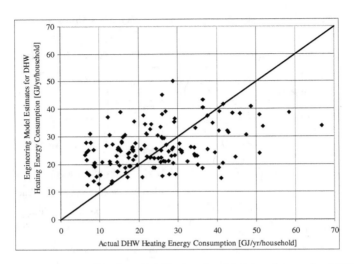

Figure 5.4.   Actual billing data and Engineering Model estimates for DHW heating energy consumption of the households in the DHW NN testing dataset

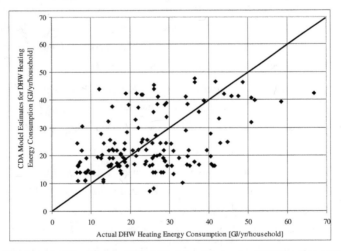

Figure 5.5. Actual billing data and CDA Model estimates for DHW heating energy consumption of the households in the DHW NN testing dataset

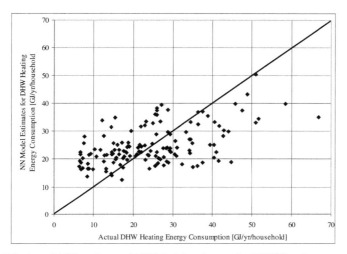

Figure 5.6. Actual billing data and NN Model estimates for DHW heating energy consumption of the households in the DHW NN testing dataset

3.2. Households without Energy Billing Data

3.2.1. Average Household DHW Heating Energy Consumption

The 1993 SHEU database has data on 8,767 households, and data on 563 of those were used to develop the DHW NN Model. There are 85 households without any DHW heating systems in the remaining 8,204 households in the 1993 SHEU database. The DHW heating prediction dataset was developed from the remaining households excluding the households using oil or wood fuelled, or tankless DHW heating systems, since these households were not represented in the DHW NN training dataset which was used in the development of the DHW NN Model. Thus, the DHW heating energy consumption of the remaining 7,070 households were estimated using the NN and CDA Models. The weighted average DHW heating energy consumption estimated by the CDA, Engineering, and NN Models for the 7,070 households, and the average percent deviations calculated using Equations 5.1a and 5.1b are given in Table 5.6. As seen in Table 5.6, the DHW heating energy consumption estimated by the CDA and the Engineering Models are 3.1% and 4.5% lower than the NN Model estimate, respectively.

112

Table 5.6. Weighted average DHW heating energy consumption estimated by the CDA, Engineering, and NN Models for the 7,070 households and average percent deviations

Model	Weighted Average DHW Heating Energy Consumption [GJ/yr/household]	Average Deviation [%]
NN	26	-
CDA	25	-3.1
Engineering	25	-4.5

The OEE estimated the DHW heating end-use energy consumption in 1993 to be 284 PJ (OEE, 2002), while the total number of households in 1993 was estimated to be 10,359,216 by Statistics Canada (1993). Therefore, the average household DHW heating energy consumption in 1993 is 27 GJ/yr/household. As seen in Table 5.6, the NN Model DHW heating energy consumption estimate is the closest to the estimate by OEE.

3.2.2. DHW Heating Energy Consumption Categorized Based on Energy Source and Fuel Type

The weighted average DHW heating energy consumption of the 7,070 households categorized based on the energy source and fuel type used for DHW heating is given in Table 5.7. As seen in Table 5.7, the households using electricity for DHW heating have lower DHW heating energy consumption than those using natural gas or propane. This is largely due to the fact that the end-use efficiency of the DHW heating systems using electricity is higher than the DHW heating systems using natural gas or propane. As it was given in Table 4.14, the end-use efficiency of the DHW heating systems using electricity is 82.4%, and for the natural gas or propane systems this figure reduces to 55.4%. Therefore, the households using natural gas or propane would have higher DHW heating energy consumption values than the households using electricity, even if both households would have the same DHW consumption and average annual ground temperatures.

Table 5.7. Weighted average DHW heating energy consumption of the 7,070 households based on the DHW heating energy source and fuel type

DHW Energy Source and Fuel Type	Weighted Average DHW Heating Energy Consumption [GJ/yr/household]		
	NN Model	CDA Model	Engineering Model
Electricity	21	19	21
Natural Gas	32	33	30
Propane	31	37	33

These figures show that the NN Model, together with the CDA and the Engineering Models, is capable of capturing the difference in the DHW heating energy consumption of the households using natural gas, propane, and electricity.

4. SH Energy Consumption

4.1. Households with Energy Billing Data

The prediction performance of the NN, CDA, and Engineering Models was assessed by comparing the estimates of the models with actual energy consumption data from the 307 households in the SH NN testing dataset. The results are presented in Table 5.8. As seen from Table 5.8, the CDA and the Engineering Models have lower R^2 and higher CV values than the NN Model, which shows that the CDA and the Engineering Models have a lower prediction performance than the NN Model.

Table 5.8. Prediction performance of the NN, CDA, and Engineering Models- SH

Model	R^2	CV
Engineering Model	0.778	2.877
CDA Model	0.892	2.007
NN Model	0.908	1.871

The estimates of the Engineering, CDA, and NN Models are plotted along with the actual energy consumption data for the 307 households in the SH NN testing dataset in Figures 5.7 to 5.9.

114

The SH NN Model, as well as, the Engineering and the CDA Models, were unable to accurately predict the SH energy consumption of most of the households with consumption values lower than 30 GJ/yr as seen in Figures 5.7 to 5.9. When the input data for these households were analyzed, it was found that these households use electricity for SH. As stated in Section 2.3.1 of Chapter 4, the amount of electricity consumed in these households for SH was calculated by subtracting the ALC and DHW heating electricity consumption predicted by the ALC and DHW NN Models from the total billed electricity consumption. Therefore, the annual SH electricity consumption of these households contain the cumulative errors from the ALC and DHW NN Models.

When the input units of these households were examined, it was found that based on the values of the input units these households could not be expected to have such low SH energy consumption values. Most of these households are single-detached, located at areas with HDD values higher than 4,500 °C-day, and have average wall areas. This shows that there are other factors affecting the SH energy consumption than those reported in the 1993 SHEU database and represented by the input units in the model (such as long vacations in winter months) that would decrease SH energy consumption of the households.

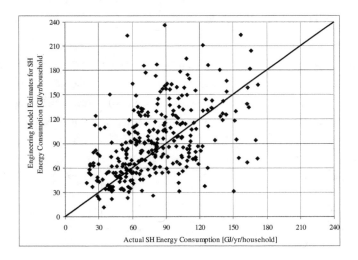

Figure 5.7.   Actual billing data and Engineering Model estimates for SH energy consumption of the households in the SH NN testing dataset

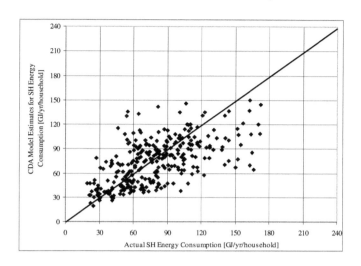

Figure 5.8.   Actual billing data and CDA Model estimates for SH energy consumption of the households in the SH NN testing dataset

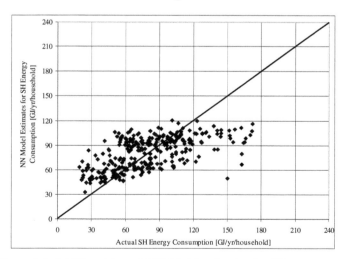

Figure 5.9.   Actual billing data and NN Model estimates for SH energy consumption of the households in the SH NN testing dataset

4.2. Households without Energy Billing Data

4.2.1. Average Household SH Energy Consumption

The 1993 SHEU database has data on 8,767 households, and data on 1,228 of those were used to develop the SH NN Model. Out of the 7,439 remaining households, 987 use wood and 153 use heat pump SH systems. Since these households were not represented in the SH NN training dataset used in the development of the SH NN Model, they were removed from further analysis, and the SH energy consumption of the remaining 6,399 households were estimated using the NN and the CDA Models. The weighted average SH energy consumption estimated by the CDA, Engineering, and NN Models for the 6,399 households, and the average percent deviations calculated using Equations 5.1a and 5.1b are given in 5.9. As seen in Table 5.9, the SH energy consumption estimated by the CDA and the Engineering Models are 3.4% and 6.9% lower than the NN Model estimate, respectively.

Table 5.9. Weighted average SH energy consumption estimated by the CDA, Engineering, and NN Models for the 6,399 households and average percent deviations

Model	Weighted Average SH Energy Consumption [GJ/yr/household]	Average Deviation [%]
NN	80	-
CDA	75	-6.9
Engineering	77	-3.4

The OEE estimated the SH end-use energy consumption in 1993 to be 837 PJ (OEE, 2002), while the total number of households in 1993 was estimated to be 10,359,216 by Statistics Canada (1993). Therefore, the average household SH energy consumption in 1993 is 81 GJ/yr/household. As seen in Table 5.9, the NN Model SH energy consumption estimate is the closest to the estimate by OEE.

4.2.2. SH Energy Consumption Categorized Based on Energy Source and Fuel Type, Dwelling Type and Age Category, and Wall Area

The weighted average SH energy consumption of the 6,399 households categorized based on the energy source and fuel type and end-use efficiency of the SH equipment is given in Table 5.10. As seen in Table 5.10, the weighted average SH energy consumption increases as the end-use efficiency of the SH equipment decreases. As expected, the NN Model SH energy consumption estimates increase smoothly as the end-use efficiency of the SH equipment decreases, whereas the CDA and Engineering Models appear to overestimate the reduction in end-use energy consumption as a result of using 100% efficient electric heating instead of high efficiency fossil fuel furnaces.

Table 5.10. Weighted average SH energy consumption of the 6,399 households based on energy source and fuel type and end-use efficiency of SH equipment

SH Energy Source and Fuel Type	End-use Efficiency [%]	Weighted Average SH Energy Consumption [GJ/yr/household]		
		NN Model	CDA Model	Engineering Model
Electricity	100	58	40	53
Oil, Natural Gas, Propane	High (90% or higher)	68	81	71
	Medium (80-75%)	89	91	91
	Standard (65-50%)	98	92	93

The weighted average SH energy consumption of the 6,399 households based on dwelling type, dwelling age category, and wall area are given in Tables 5.11 to 5.13, respectively. As given in Table 5.11, the weighted average SH energy consumption estimates of all three models for single-detached dwellings are higher than that of single-attached dwellings. This is a reasonable outcome since a single detached dwelling has more exposed envelope area compared to a similarly sized single attached dwelling, and consequently, requires more energy for space heating.

Table 5.11. Weighted average SH energy consumption of the 6,399 households based on dwelling type

Dwelling Type	Weighted Average SH Energy Consumption [GJ/yr/household]		
	NN Model	CDA Model	Engineering Model
Single-attached	67	55	61
Single-detached	84	80	82

The weighted average SH energy consumption estimates of the models decrease as the dwellings get younger, and increase as the wall area of the dwellings increase, as seen in Table 5.12 and 5.13, respectively. These findings are also in agreement with expectations. The envelope integrity and insulation levels of newer buildings are higher than older ones, resulting in less energy consumption in newer residences; and wall area is directly proportional with envelope heat loss.

Table 5.12. Weighted average SH energy consumption of the 6,399 households based on dwelling age category

Year Category	Weighted Average SH Energy Consumption [GJ/yr/household]		
	NN Model	CDA Model	Engineering Model
1. Before 1941	90	82	98
2. 1941 – 1960	85	77	80
3. 1961 – 1977	82	78	74
4. 1978 – 1982	75	69	70
5. 1983 – 1988	63	59	59
6. 1989 or later	62	64	63

Table 5.13. Weighted average SH energy consumption of the 6,399 households based on wall area of the dwelling

Wall Area [m^2]	Weighted Average SH Energy Consumption [GJ/yr/household]		
	NN Model	CDA Model	Engineering Model
Less than 100	75	61	58
100 – 130	80	72	82
More than 130	84	85	85

119

These figures show that the SH NN, CDA, and the Engineering Models are all capable of capturing the differences in the SH energy consumption of households with different SH equipment efficiencies, wall areas, and dwelling types and ages.

5. Household Energy Consumption

The DHW heating and SH energy consumption of 2,096 households in the 1993 SHEU database could not be estimated by the DHW and SH NN Models due to insufficient data. The household energy consumption of the remaining 6,671 households in the 1993 SHEU database were computed by combining the ALC, DHW heating, and SH energy consumption estimates of the NN and CDA Models. Table 5.14 gives the weighted average household energy consumption estimates of the NN, CDA, and Engineering Models, and the average percent deviations calculated using Equations 5.1a and 5.1b. As seen in Table 5.14, the household energy consumption estimated by the CDA and the Engineering Models are 5% and 2.8% lower than the NN Model estimate, respectively.

Table 5.14. Weighted average household energy consumption of the 6,671 households in the 1993 SHEU database and average percent deviations

Model	Weighted Average Household Energy Consumption [GJ/yr/household]	Average Deviation [%]
NN	139	-
CDA	132	-5.0
Engineering	135	-2.8

The OEE estimated the total household energy consumption in 1993 to be 1,386 PJ (OEE, 2002), while the total number of households in 1993 was estimated to be 10,359,216 by Statistics Canada (1993). Therefore, the average total household energy consumption in 1993 is 134 GJ/yr/household. As seen in Table 5.14, the estimates by all three models are close to the estimate by OEE.

5.1. Household Energy Consumption Categorized Based on SH Energy Source and Fuel Type, Dwelling Type and Age Category, and Province

The household energy consumption of the 6,671 households estimated by the NN, CDA, and the Engineering Models were categorized based on SH energy source and fuel type, dwelling type and age category. The results are presented in Tables 5.15 to 5.17.

Dwellings that are electrically heated, single attached, and built after 1988 have the lowest average weighted household energy consumption according to the NN, CDA, and Engineering Models estimates, as seen in Tables 5.15 to 5.17. Since, the SH end-use energy consumption accounts for about 60% of the total household energy consumption (OEE, 2002), trends similar to the SH energy consumption estimates given in Tables 5.10 to 5.12 are seen in Tables 5.15 to 5.17.

The CDA Model seem to estimate the energy consumption of the electrical heated households lower than the NN Model as seen in Table 5.15, similar to the estimates given in Table 5.10.

Since, the SH end-use energy consumption is a major component of the household energy consumption, single detached dwellings have higher energy consumption estimates as seen in Table 5.16. The estimates of all three models for the single detached dwellings are higher than the ones for single detached dwellings.

Table 5.15. Weighted average household energy consumption of the 6,671 households based on SH energy source and fuel type

SH Energy Source and Fuel Type	Weighted Average Household Energy Consumption [GJ/yr/household]		
	NN Model	CDA Model	Engineering Model
Electricity	107	89	107
Natural Gas	155	155	150
Oil	145	132	137
Propane	142	147	153

Table 5.16. Weighted average household energy consumption of the 6,671 households based on dwelling type

Dwelling Type	Weighted Average Household Energy Consumption [GJ/yr/household]		
	NN Model	CDA Model	Engineering Model
Single-attached	120	99	112
Single-detached	144	140	141

As seen in Table 5.17, the households built after 1988 have the lowest average weighted household energy consumption estimates. There is not a smooth decrease in the household energy consumption estimates as the dwellings get younger, as seen in the SH estimates in Table 5.12. This is largely due to the differences in the distribution of dwelling types, wall areas, and energy sources and fuel types in each age category of households.

Table 5.17. Weighted average household energy consumption of the 6,671 households based on dwelling age category

Year Category	Weighted Average Household Energy Consumption [GJ/yr/household]		
	NN Model	CDA Model	Engineering Model
1. Before 1941	141	135	149
2. 1941 – 1960	141	131	135
3. 1961 – 1977	144	135	135
4. 1978 – 1982	141	133	132
5. 1983 – 1988	128	127	124
6. 1989 or later	121	120	124

The average household energy consumption in each province was calculated using the estimates of the NN, CDA, and Engineering Models. The results are presented in Figure 5.10. As seen in Figure 5.10, the estimates of the three models are in agreement. The average household energy consumption in Quebec is found to be the lowest, whereas Alberta and Saskatchewan have the highest household energy consumption.

SH energy consumption accounts for about 60% of the total household energy consumption. Therefore, factors such as end-use efficiency, and energy source and fuel type of the SH equipment have significant effects on the total household energy consumption. Consequently, the trend seen in Figure 5.10 is mainly due to the fact that, in the 1993 SHEU database, 79% of the households in Quebec, and, respectively, 1% and 5% of the households in Alberta and Saskatchewan, have electrical SH equipment that have 100% end-use efficiency. In addition, 66% and 69% of the households in Alberta and Saskatchewan, respectively, have standard (50-65%) efficiency natural gas, oil, or propane fueled SH equipment. These explain the high household energy consumption trends in Alberta and Saskatchewan.

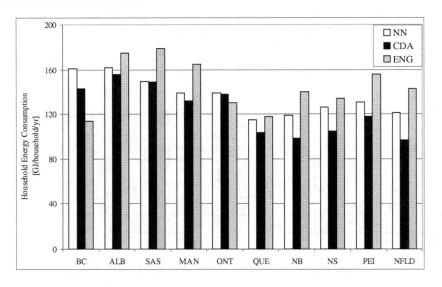

Figure 5.10. NN, CDA, and Engineering Models provincial household energy distribution

6. Assessment of Socio-economic Factors

The effects of the socio-economic factors, such as income, dwelling ownership, size of area of residence, on the ALC, DHW heating, and SH energy consumption of the households in the 1993 SHEU database were studied using the NN and CDA Models, and the results are presented in the following sections.

6.1. Impact of Socio-economic Factors on ALC Energy Consumption

The input dataset of the ALC NN Model contains information on the socio-economic characteristics of the households; such as household income (in $10,000 increments), dwelling ownership (rent or own), size of area of residence, dwelling type (single attached or single detached), and number of children and adults. The CDA Electricity Model given in Equation 4.28 includes the household income and the number of occupants as socio-economic factors on ALC electricity consumption. All other socio-economic factors that were initially included in the CDA EM equation (4.28) were eliminated as a result of statistical significance and multicollinearity problems as explained in Section 3.1.3 of Chapter 4. Therefore, the capability of the CDA Model to evaluate the effects of a large number of socio-economic factors on the ALC energy consumption is significantly reduced.

The effects of the socio-economic factors on the ALC electricity consumption were assessed using the NN and CDA Models. The results are plotted in Figures 5.11 and 5.17, respectively.

As seen from Figures 5.11 and 5.12, ALC electricity consumption of the households varies as follows:

- ALC electricity consumption of households increase with income. This is due to the fact that as income increases, number of appliances, living area, number of lights, and space cooling equipment load increase, and, consequently, the ALC electricity consumption increases. It is interesting to note that at low-income levels, the slope of the NN Model curve becomes less, indicating that the effect of income on the ALC electricity consumption is lower. In other words, as income decreases, the ALC energy consumption reaches a minimum and stays constant. On the other hand, the CDA Model income estimates form a straight line indicating that ALC electricity consumption decreases with a constant rate (*i.e.* with a slope of 0.0064) as income decreases, and does not stay constant at low income levels.

- NN Model ALC electricity consumption estimate of a single detached dwelling is higher than that of a single attached dwelling. This is due to the fact that single detached dwellings have larger living areas than single attached ones. As living area increases, the number of lights in the dwelling increases, as well as the space cooling load increases.

- NN Model ALC electricity consumption estimate of an owner occupied household is higher than that of a renter occupied household. In the 1993 SHEU database, the majority of the renter occupied households are single attached. As stated above, single attached dwellings have smaller living areas; consequently, they have lower ALC energy consumption.

124

- NN Model ALC electricity consumption estimate of a household decreases as the population of the area increases. This is due to the fact that almost all of the households located in areas with population less than 15,000 are single detached, and have larger living areas and more outdoor lamps than the single attached dwellings located at populated areas. Therefore, as the area and the number of lights increase, the ALC energy consumption increases.
- ALC electricity consumption of a household increases as the number of children and adults increase. The increase in the number of adults has a more significant influence on the ALC energy consumption than the increase due to the number of children. As the number of adults increases, the number of TVs, VCRs, stereos, and computers increase; also, the number of bedrooms increases, increasing the living area. However, the increase in the number of children does not generally increase the number of appliances and bedrooms. Therefore, the effect of an increase in the number of adults on the ALC energy consumption is more significant than the increase due to the increase in the number of children. The CDA Model estimates show a substantially higher increase in the ALC electricity consumption as the number of children increases, whereas the estimated effect of the increase in the number of adults is of similar magnitude for both models.

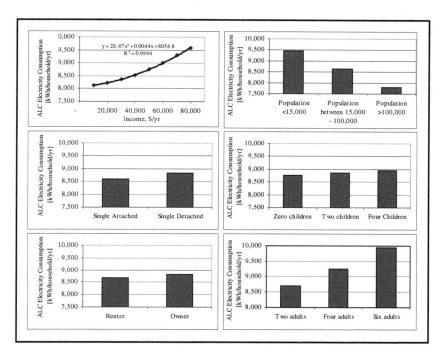

Figure 5.11. Effects of socio-economic factors on the ALC energy consumption of the households estimated by the NN Model

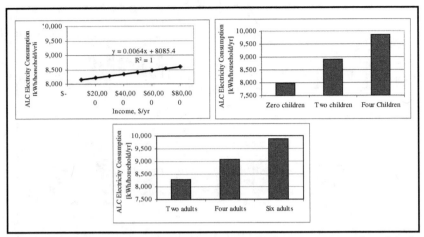

Figure 5.12. Effects of socio-economic factors on the ALC energy consumption of the households estimated by the CDA Model

6.2. Impact of Socio-economic Factors on DHW Heating Energy Consumption

The socio-economic factors included in the input dataset of the DHW NN Model are household income, dwelling type and ownership, and size of area of residence. The only socio-economic factor affecting the estimation of the DHW heating energy consumption in the CDA Model is the dwelling type of the households that use natural gas for DHW heating as seen in the CDA NGM given in Equation 4.36. All other socio-economic factors that were initially included in the CDA equations were eliminated as a result of statistical significance and multicollinearity problems as explained in Sections 3.1.3, 3.2.3, and 3.3.3 of Chapter 4. Thus, the capability of the CDA Model to evaluate the effects of socio-economic factors is significantly reduced.

The effects of the socio-economic factors on DHW heating energy consumption were examined and the results obtained from the NN Model are plotted in Figure 5.13. As seen from Figure 5.13, the estimated DHW heating energy consumption of the households varies as follows:

- DHW heating energy consumption increases linearly with a slope of 0.00004, as the income of the household increases.

- The average DHW heating energy consumption of a single detached dwelling is higher than that of a single attached dwelling. This is due to the fact that in the 1993 SHEU database as the number of occupants increases, the living area of the dwellings increases. As mentioned in

Section 6.1 of Chapter 5, single detached dwellings have larger living areas than single attached ones. Consequently, the number of occupants is higher in single detached dwellings than in single attached ones, so is the DHW heating energy consumption.

- DHW heating energy consumption of an owner occupied household is higher than that of a renter occupied household. As mentioned in Section 6.1 of Chapter 5, the majority of the renter occupied households are single attached. Consequently, the single attached dwellings have smaller living areas, fewer number of occupants, and lower DHW heating energy consumption.
- DHW heating energy consumption decreases as the population of the area increases. As mentioned in Section 6.1 of Chapter 5, almost all of the households located in areas with population less than 15,000 are single detached, and have larger living areas, and consequently, higher number of occupants than the single attached dwellings located at populated areas. Therefore, as the number of occupants increases, DHW heating energy consumption increases.

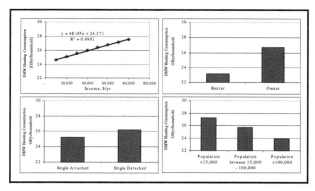

Figure 5.13. Effects of socio-economic factors on the DHW heating energy consumption of the households estimated by the NN Model

Natural gas consumption of single attached and single detached dwellings for DHW heating were estimated using the CDA NGM to be 11.32 GJ/yr/household and 38.18 GJ/yr/household. This corresponds to a difference of 27 GJ/yr/household, which is unacceptably high since the average natural gas consumption for DHW heating was estimated to be 33 GJ/yr/household using the CDA NGM, as given in Table 5.7, and the dwelling type can not have such a large impact on DHW heating energy consumption. On the other hand, the difference in the DHW heating natural gas consumption of the single attached and single detached dwellings was estimated using the NN

128

Model to be 0.6 GJ/yr/household.

6.3. Impact of Socio-economic Factors on SH Energy Consumption

The input dataset of the SH NN Model contains information on the socio-economic characteristics of the households; such as household income, dwelling ownership, size of area of residence, and number of children and adults. The effects of the socio-economic factors on the SH energy consumption were examined, and the results obtained from the NN Model are plotted in Figure 5.14. As seen from Figure 5.14, SH energy consumption estimates vary as follows:

- SH energy consumption increases as income increases. This is due to the fact that households with higher income levels have larger dwellings. As the area of the dwelling increases, SH energy consumption increases linearly with a slope of 0.0001.

- SH energy consumption of an owner occupied household is higher than that of a renter occupied household. As mentioned in Sections 6.1 and 6.2 of Chapter 5, majority of the owner occupied dwellings are single detached, and the SH energy consumption of the single detached dwellings is higher than the single attached ones, as given in Table 5.11. Consequently, the SH energy consumption of owner occupied households is higher than renter occupied ones.

- SH energy consumption of a household decreases as the population of the area increases. As mentioned in Sections 6.1 and 6.2 of Chapter 5, almost all of the households located in areas with population less than 15,000 are single detached, and as given in Table 5.11, the SH energy consumption of single detached dwellings is higher than single attached ones. Therefore, SH energy consumption of the households located in rural areas is higher than those located in urban areas.

- SH energy consumption of a household increases as the number of children and adults increase. This is due to the fact that as the number of occupants increases, the living area increases, as does the SH energy consumption.

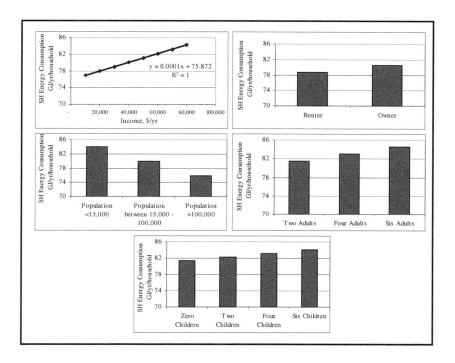

Figure 5.14. Effects of socio-economic factors on the SH energy consumption of the households estimated by the NN Model

The CDA EM given in Equation 4.28 includes the household income, and the CDA NGM given in Equation 4.36 includes the number of adults as socio-economic factors on the SH energy consumption. All other socio-economic factors that were initially included in the CDA equations were eliminated as a result of statistical significance and multicollinearity problems as explained in Sections 3.1.3, 3.2.3, and 3.3.3 of Chapter 4. Therefore, only household income and number of adults could be evaluated using the CDA Model. The results from the CDA EM and NGM are plotted in Figure 5.15. For comparison purposes, the effect of income on electrical heated households, and the effect of number of adults on natural gas heated households were estimated using the SH NN Model, and results are plotted in Figure 5.16.

As shown in Figures 5.15 and 5.16, as the income of the household increases, the electricity consumption for SH increases in a linear fashion. This is similar to the NN estimates given in Figure 5.14. However, the slope of the line plotted using the NN Model estimates is about one third of the one from the CDA EM.

130

As shown in Figures 5.15 and 5.16, as the number of adults increases, the natural gas consumption for SH increases, similar to the NN estimates given in Figure 5.14. Compared to NN SH Model estimates, the CDA NGM estimates show a substantially higher increase in the natural gas consumption for SH as the number of adults increases.

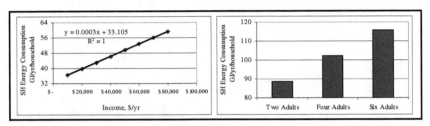

Figure 5.15. Effect of income on electrical SH energy consumption and effect of number of adults on natural gas SH energy consumption estimated by the CDA Model

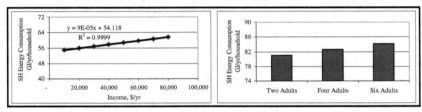

Figure 5.16. Effect of income on electrical SH energy consumption and effect of number of adults on natural gas SH energy consumption estimated by the SH NN Model

7. Closing Remarks

This chapter presents the results of the comparative assessment of the NN, CDA, and Engineering Models. The detailed information on the development of the NN and the CDA Models was given in Chapter 4. The Engineering Model was developed by Farahbakhsh (1997) and Farahbakhsh *et al.* (1997, 1998) using the 1993 SHEU database.

In the first section of the chapter, the prediction performances of the models were compared based on the actual and estimated end-use energy consumption of the households in the 1993 SHEU database. The prediction performance of the NN Model was found to be higher than those of the CDA and the Engineering Models.

The weighted average end-use energy consumption of the households without billing data

131

was estimated by the models, and the average percent deviations between the estimates of the NN Model, and the CDA and Engineering Models were calculated for each end-use. The weighted average end-use energy consumption estimates of the models were found to be close to each other with percent deviations in the range of -6.9 to 2.5. The end-use estimates by the three models were also close to the estimates by NRCan and OEE.

The central A/C unit, second refrigerator, and the furnace fan/boiler pump electricity consumption were successfully estimated using the ALC NN Model, whereas the NN Model failed to accurately estimate the electricity consumption of appliances with high saturation values.

The weighted average DHW and SH end-use energy consumption of the households without billing data were categorized based on end-use energy source and fuel type, dwelling type and age. It was found that:

– The DHW NN Model, as well as the CDA and the Engineering Models, are capable of capturing the difference in the DHW heating energy consumption of households using DHW heating equipment with different efficiencies.

– The SH NN Model, together with the CDA and the Engineering Models, are capable of capturing the differences in the SH energy consumption of the households with different SH equipment efficiencies, wall areas, and dwelling types and ages.

In the second section of the chapter, the total energy consumption of the households in the 1993 SHEU database were computed by combining the ALC, DHW heating, and SH energy consumption estimates of the NN, CDA, and Engineering Models. The weighted average household energy consumption estimated by the NN, CDA, and Engineering Models were 139, 132, and 135 GJ/yr/household, respectively. The estimates of all three models are close to the OEE estimate of 134 GJ/yr/household.

The categorized household energy consumption estimates by all three models indicate that electrically heated, and single attached dwellings, and dwellings built after 1988 have the lowest average weighted household energy consumption. The provincial distribution of the household end-use energy consumption estimates demonstrated that households in Quebec have lower; and households in Alberta and Saskatchewan have higher end-use energy consumption. This is due to the fact that the majority of the households in Quebec, and minority of the households in Alberta and Saskatchewan have electrical SH equipment, which have 100% end-use efficiency. Also more than half of the households in Alberta and Saskatchewan have standard (50-65%) efficiency natural gas, oil, or propane fueled SH equipment resulting in high household end-use energy consumption trends in Alberta and Saskatchewan.

In the last section of the chapter, the effects of the socio-economic factors on the ALC, DHW heating, and SH end-use energy consumption of the households in the 1993 SHEU database were assessed by the NN and the CDA Models. It was found that end-use energy consumption of households increases as the household income, and the number of adults and children increase, and decreases as the population of the area increases. It was also found that the end-use energy consumption of single detached dwellings are higher than that of single attached ones, and owner occupied households have higher end-use energy consumption than renter-occupied ones.

The capability of the CDA Model to evaluate the effects of the socio-economic factors is significantly lower than the NN Model since majority of the socio-economic factors were excluded from the model equation to develop a model free of insignificant variables and multicollinearity. However, the NN Model has the capability to evaluate the effects of the socio-economic factors, which are included in the model. The effects of the socio-economic factors cannot be evaluated by the Engineering Model since these factors were not included into the model. However, if sufficient data on socio-economic factors is available, the Engineering Model can evaluate the effects of socio-economic factors when they are included into the model in the form of occupancy profiles or preference profiles.

The comparisons of the model estimates show that the NN Model has a higher prediction performance than the CDA and the Engineering Models, can estimate the energy consumption of individual appliances, can successfully evaluate the differences in end-use and total household energy consumption based on various categories, and has the capability to evaluate the effects of a large number of socio-economic factors.

In the following chapter, impacts of energy saving measures on DHW and SH energy consumption are assessed using the NN Model.

Chapter 6

Assessment of Energy Saving Scenarios

1. Overview

In this chapter, impacts of various energy saving scenarios on DHW and SH energy consumption of the households in the 1993 SHEU database are assessed. The analysis was conducted using the NN Model, and the results are compared with those obtained using the Engineering Model, and some other studies, where available. The CDA Model does not contain any of the variables used in the DHW and SH energy saving scenarios, thus it was not possible to evaluate energy saving scenarios using the CDA Model. The CDA Model is not suitable to conduct such assessments due to the limited number of variables the model can accommodate.

To assess the impacts of a wide variety of energy saving measures using the NN Model, the households that undertake the energy saving measures should be well represented in the training datasets. On the other hand, the engineering method based models have significantly higher level of flexibility in evaluating energy conservation measures, regardless of the number of households in the datasets.

The NN Model consists of three networks, and each one of these networks estimates a single end-use energy consumption. As stated in Section 2 of Chapter 1, the energy efficiency measures have complex interrelated effects on the end-use energy consumption, *e.g.* the increase in the lighting efficiency increases the SH energy consumption. However, the interrelated effects due to energy efficiency measures on end-uses cannot be evaluated using the NN Model, since each end-use is predicted separately. On the other hand, the Engineering Model can evaluate the effects of energy efficiency measures on each end-use, since energy consumption is estimated using

134

thermodynamics and heat transfer principles.

2. DHW Heating Energy Consumption

Three energy saving scenarios evaluated using the DHW NN Model are insulating the hot water pipes and increasing the end-use efficiency of DHW heating systems. The estimated reduction in the DHW heating energy consumption for each energy saving scenario is given in the following sections.

2.1. Insulating Hot Water Pipes

The number of households that do not have insulation around hot water pipes is 5,767 in the DHW prediction dataset of 7,070 households. There is a weighted average DHW heating energy consumption difference of 0.7 GJ's between the households with and without insulation around hot water pipes. In order to examine the effect of adding insulation around hot water pipes, data on the 5,767 households were modified to reflect the addition of insulation around hot water pipes. The DHW NN Model was used to estimate the DHW heating energy consumption of these 5,767 households. Thus, as shown in Table 6.1, the difference between the predicted DHW heating energy consumption of the 5,767 households with and without insulation around hot water pipes provides the magnitude of the effect of insulating hot water pipes on the DHW heating energy consumption. The average reduction in the DHW heating energy consumption by adding insulation around the hot water pipes is estimated to be 0.9 GJ/yr/household.

Table 6.1. Average DHW heating energy consumption of 5,767 households in set of 7,070 households with and without insulation around hot water pipes

	Average DHW Heating Energy Consumption [GJ/yr/household]
5,767 Households without insulation around the hot water pipes	26.2
5,767 Households with insulation around the hot water pipes	25.3
Reduction in the DHW heating consumption	0.9

In its present form, the Engineering Model does not address the presence of insulation around hot water pipes, therefore, the results obtained from the NN Model were compared with the results of a study conducted by Pontikakis and Ruth (1999). The authors estimated a reduction of 1.5-1.7 % in the heat loss from hot water pipes due to insulation, which correspond to a reduction in the DHW heating energy consumption of about 0.5 GJ/yr/household. The DHW NN model estimate for the reduction in the DHW heating energy consumption due to the insulation of hot water pipes is about two times higher than Pontikakis and Ruth's estimate.

The reason for this over prediction could be the low percentage of households with hot water pipe insulation in the DHW NN training dataset: only 16% of households in the DHW NN training dataset had insulation around their hot water pipes.

2.2. Increasing the End-use Efficiency of DHW Heating Systems

There are 2,765 households using natural gas or propane heated DHW systems in the DHW prediction dataset of 7,070 households. These systems were assumed to have an end-use efficiency of 55.4%. In order to examine the effect of upgrading the DHW heating system end-use efficiency, a dataset was developed by increasing the end-use efficiency of natural gas or propane heated DHW systems from 55.4% to 65% in the input files of the 2,765 households. The DHW NN Model was used to estimate the DHW heating energy consumption of these 2,765 households. Thus, as shown in Table 6.2, the difference between the predicted energy consumption of 2,765 households with an end-use efficiency of 55.4% and 65% provides the magnitude of the effect of end-use efficiency increase on the DHW heating energy consumption, which is 3.9 GJ/yr/household. A reduction of 4.5 GJ/yr/household was also estimated using the Engineering Model for the same end-use efficiency improvement. Thus, the DHW NN model and the Engineering Model estimates are in good agreement.

Table 6.2. Average DHW heating energy consumption of 2,765 households in set of 7,070 households at two different DHW system efficiencies

	Average DHW Heating Energy Consumption [GJ/yr/household]
2,765 Households with an end-use efficiency of 55.4%	31.9
2,765 Households with an end-use efficiency of 65%	28.0
Reduction in the DHW heating consumption	3.9

3. SH Energy Consumption

The energy saving scenarios evaluated using the SH NN Model are upgrading the glazing of the windows and end-use efficiency of the SH equipment, and lowering the overnight temperature. Other scenarios could not be studied due to the poor representation of the households that undertake the scenarios in the SH NN training dataset, as mentioned in Section 1 of Chapter 6. The reduction in the SH energy consumption associated with each energy saving scenario is given in the following sections.

3.1. Upgrading the Glazing of the Windows

The SH prediction dataset contains 1,157, 6,128, and 6,200 households, respectively, with single, double, and both types of glazed windows. In order to examine the effects of upgrading the glazing of the windows, households with single glazed windows were upgraded to double and triple glazed windows, while those with double glazed windows were upgraded to triple glazed windows. The SH NN Model was used to estimate the SH energy consumption of the households with glazing upgrades. Thus, the difference between the predicted SH energy consumption of the households before and after the upgrades provides the magnitude of the effect of window glazing upgrade on SH energy consumption.

Table 6.3 gives the average SH energy consumption of the households before and after upgrades to their window glazing. The SH energy consumption decreases due to window glazing upgrades, with the exception of the scenario in which single glazed windows were upgraded to double glazed windows. The reason for this anomaly could be the low percentage of households with single glazed windows in the SH NN training dataset: only 14% of the households in the SH NN training dataset have single glazed windows. Thus, the reduction in the SH energy consumption of the households due to single glazed window upgrades cannot be accurately captured by the SH NN Model.

The SH energy consumption reduction due to the upgrade from single to triple glazed windows is 2 GJ/yr/household, which is significantly small compared to the reduction of 13.5 GJ/yr/household estimated by Guler *et al.* (2001) using the Engineering Model. The reduction estimated by Guler *et al.* (2001) as a result of upgrading single glazed windows to double glazed

137

was about 9.8 GJ/yr/household. Another study conducted for US Department of Energy estimated a reduction of 8.9 GJ/yr/household by upgrading all single glazed windows to double glazed windows (Conachen, 2001). These results indicate that the SH NN Model fails to predict the effect of the energy saving scenarios based on single glazed window upgrades. This is due to the low number of households with single glazed windows in the SH NN training dataset.

Table 6.3. Average SH energy consumption of the households before and after window glazing upgrades

Upgrade	Number of Households	Average SH Energy Consumption [GJ/yr/household]		
		Before	After	Difference
Single to double glazed	1,157	84.4	84.8	-0.4
Single to triple glazed	1,157	84.4	82.6	1.8
Double to triple glazed	6,128	80.0	76.4	3.6
Single to double glazed Double to triple glazed	6,200	80.2	76.7	3.5
Single to triple glazed Double to triple glazed	6,200	80.2	76.2	4.0

On the other hand, the reduction of 3.6 GJ/year/household due to upgrading double glazed windows to triple glazed estimated using the SH NN Model is close to the 4.3 GJ/yr/household estimated by Guler *et al.* (2001) using the Engineering Model. This indicates that the effect of upgrading double glazed windows can be successfully predicted by the SH NN Model.

The number of households with both single and double glazed windows is 6,200 in the prediction dataset. The reduction in SH energy consumption estimated using the SH NN Model as a result of upgrading single glazed windows to double glazed, and double glazed windows to triple glazed is 3.5 GJ/yr/household, while upgrading both single and double glazed windows to triple glazed is 4 GJ/yr/household. These estimates are lower than expected. The reason for this underestimation is likely due to the afore mentioned inability of the SH NN Model to accurately predict the savings due to upgrading single glazed windows.

3.2. Upgrading SH Equipment

The number of households with standard and medium efficiency SH equipment in the SH prediction dataset are 2,350 and 1,327, respectively. In order to examine the effects of upgrading SH equipment, all standard efficiency SH equipment were replaced with medium and high efficiency SH equipment, and medium efficiency SH equipment were replaced with high efficiency equipment. The SH NN Model was used to estimate the SH energy consumption of the households with end-use efficiency upgrades. Thus, the difference between the predicted SH energy consumption of the households with and without SH equipment end-use efficiency upgrades provides the magnitude of the effect of SH equipment end-use efficiency upgrade on SH energy consumption. Table 6.4 gives the average SH energy consumption of the households before and after SH equipment end-use efficiency upgrades. The SH energy consumption of the households using standard and medium efficiency SH equipment decreases significantly as their SH equipment are upgraded to medium and high efficiencies.

Table 6.4. Average SH energy consumption of the households before and after SH equipment end-use efficiency upgrades

Upgrade	Number of Households	Average SH Energy Consumption [GJ/yr/household]		
		Before	After	Difference
Standard to medium efficiency	2,350	97.6	87.1	10.5
Standard to high efficiency	2,350	97.6	67.6	30.0
Medium to high efficiency	1,327	89.0	69.5	19.5
Standard to medium efficiency Medium to high efficiency	3,677	94.6	80.8	13.8
Standard to high efficiency Medium to high efficiency	3,677	94.6	68.3	26.3

The SH energy consumption of the households after the SH equipment end-use efficiency upgrades was also predicted using the Engineering Model. The percent reductions in the SH energy consumption estimated by the SH NN and Engineering Models are given in Table 6.5. The SH energy consumption percent reduction values from both models are close for each upgrade scenario.

Table 6.5. Average percent reduction in the SH energy consumption estimated by the SH NN Model and the Engineering Model

Upgrade	Average Percent Reduction in the Estimated SH Energy Consumption in 1993 per household	
	SH NN Model	Engineering Model
Standard to medium efficiency	11%	11%
Standard to high efficiency	31%	27%
Medium to high efficiency	22%	18%
Standard to medium efficiency Medium to high efficiency	15%	13%
Standard to high efficiency Medium to high efficiency	28%	24%

3.3. Night Temperature Setback

The overnight temperature of the households in the 1993 SHEU database ranges from 16°C to 24°C. There are 3,190 households in the SH prediction dataset with overnight temperatures higher than 18°C. The average overnight temperature of these households is 18.7°C. In order to examine the effect of lowering overnight temperatures, a dataset was developed by setting the overnight temperatures of these 3,190 households to 18°C. The SH NN Model was used to estimate the SH energy consumption of these 3,190 households with overnight temperature lowered to 18°C. Thus, as shown in Table 6.6 the difference between the predicted energy consumption of 3,190 households before and after setting overnight temperatures to 18°C provides the magnitude of the effect of lowering overnight temperature on SH energy consumption. The average reduction in the SH energy consumption as a result of lowering overnight temperature of the households to 18°C was 1.3 GJ/yr.

Table 6.6. Average SH energy consumption of 3,190 households before and after lowering their overnight temperatures to 18°C

	Average SH Energy Consumption [GJ/yr/household]
3,190 Households with overnight temperatures higher than 18°C	81.5
3,190 Households with overnight temperatures set to 18°C	80.2
Reduction in the SH consumption	1.3

There are, respectively, 1,203 and 4,931 households with overnight temperatures higher than 20°C and 16°C in the prediction dataset. Using the same procedure described above, the reduction in their SH energy consumption by lowering overnight temperatures to 16°C and 20°C were estimated by the SH NN Model. Figure 6.1 shows the reduction in the SH energy consumption due to lowering overnight temperatures to 16°C, 18°C, and 20°C. As seen from Figure 6.1, as the overnight temperature of households decreases, the reduction in the SH energy consumption increases.

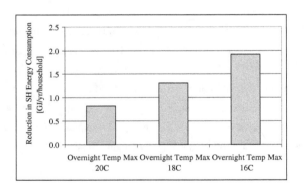

Figure 6.1.   Reduction in the SH energy consumption due to lowering overnight temperatures to 20°C, 18°C, and 16°C.

A study conducted by US Department of Energy suggested a decrease of about 5% to 15% in the annual SH energy consumption if the overnight temperatures were lowered by 6 to 8°C (EREC, 1997). The study also stated that savings from temperature setback would be greater for dwellings in milder climates than for those in more severe climates. Therefore, lowering overnight

141

temperatures by 6 to 8°C in Canadian homes which are generally located at more severe climates than US homes would provide less reduction in SH energy consumption than those stated by EREC (1997).

The average percent reduction in SH energy consumption due to lowering overnight temperature for each degree is given in Table 6.7. The average percent reduction due to lowering overnight temperature by 6°C is about 4%, which is slightly lower, as expected, than the figures reported by EREC (1997). These results indicate that the reduction in the SH energy consumption due to lowering overnight temperatures estimated by the SH NN Models agree with those reported by EREC (1997).

Table 6.7. Average percent reduction in SH energy consumption due to lowering overnight temperature for each degree

Reduction in the Overnight Temperature	Percent Reduction in the Estimated SH Energy Consumption in 1993 per household
1°C	0.7%
2°C	1.3%
3°C	2.0%
4°C	2.6%
5°C	3.2%
6°C	3.9%
7°C	4.6%
8°C	5.3%

4. Closing Remarks

This chapter presents the assessment of various energy savings scenarios conducted using the NN Model, and a comparison of the reductions estimated by the NN Model with those from other studies. The number and variety of scenarios were limited by the variables included in the NN Model. It was not possible to use the CDA model to evaluate energy saving scenarios, since the variables included in the CDA Model are limited. Thus, in terms of the capability of evaluating energy saving scenarios, the NN Model has a limited capability, while the CDA Model has none. In comparison to NN and CDA Models, the Engineering Model has a high degree of flexibility in evaluating energy saving scenarios.

The DHW energy saving scenarios evaluated using the NN Model were insulating hot water pipes and increasing the end-use efficiency of DHW heating systems. The average DHW energy consumption reduction due to addition of insulation around hot water pipes was estimated to be 0.9 GJ/yr/household. The estimated energy savings due to the addition of insulation around hot water pipes was found to be higher than the estimates of other studies. The reason for the overestimation could be the low number of households with hot water pipe insulation in the DHW training dataset.

The savings in DHW energy consumption due to increasing the end-use efficiency of natural gas/propane fuelled DHW heating systems was estimated to be 3.9 GJ/yr/household using the NN Model. This result was found to be in good agreement with the Engineering Model estimate.

The SH energy saving scenarios conducted using the NN Model were upgrading the glazing of the windows, upgrading the end-use efficiency of the SH equipment, and lowering the overnight temperature. The scenarios regarding single glazed windows upgrades resulted in lower SH energy consumption reductions than those obtained from the Engineering Model and other studies. The NN Model underestimated the energy savings because the model was not able to capture the effect of single glazed windows from a dataset of only 129 households. The amount of reduction in SH energy consumption due to upgrading double glazed windows to triple glazed was estimated to be 3.6 GJ/yr/household, which is in agreement with the Engineering Model estimate of 4.3 GJ/yr/household.

The estimated energy savings due to upgrading SH equipment end-use efficiency was found to be in good agreement with the Engineering Model estimate. The reduction in SH energy consumption due to upgrading medium efficiency SH equipment to high efficiency was estimated to be 22%, which is close to the Engineering Model estimate of 18%. The percent reduction estimated by the NN Model due to lowering overnight temperature by 6°C was about 4 %, which is close to the value stated by EREC (1997).

The comparisons in this chapter show that the NN Model can estimate the impacts of the energy savings scenarios, as long as the households that undertake the scenarios are well represented in the training dataset. However, the NN Model cannot evaluate the impact of an energy saving scenario on other energy end-uses, since each end-use is predicted separately. The Engineering Model has significantly higher level of flexibility in evaluating energy conservation measures, whereas the CDA Model is not suitable to evaluate energy efficiency measures due to the limited number of variables the model can accommodate.

143

Chapter 7

Conclusions and Recommendations

1. Conclusions

This work investigates the use of Neural Network (NN) and Conditional Demand Analysis (CDA) methods for modeling residential end-use energy consumption at the national and regional levels. In this work, end-use energy consumption models were developed for the Canadian residential sector using the NN and CDA methods, and the extensive data available in the 1993 Survey of Household Energy Use (SHEU) database of Statistics Canada (1993). Although NN's have characteristics suitable for modeling residential energy consumption at the national and regional levels, no NN based model had been reported in the literature before the current work. Similarly, the CDA method had not been used to model residential energy consumption at the national level, although there are several studies where CDA was used to model energy consumption at the regional level. Thus, the NN and CDA Models developed in this work are the first of their kind, and represent original contributions to the state-of-the-knowledge in energy modeling. Parts of this work have already been published in peer-reviewed open literature (Aydinalp *et al.*, 2002a; Aydinalp *et al.*, 2002b; Aydinalp *et al.*, 2001a; Aydinalp *et al.*, 2001b; Aydinalp *et al.*, 2000).

The NN Model of the residential sector developed in this work consists of three independent networks (sub-models), each of which predicts the energy consumption for three major energy end-uses in a household. These three sub-models are: (1) Appliance, lighting and cooling (ALC) model, (2) Domestic hot water (DHW) heating model, (3) Space heating model. These three models together make up the overall end-use energy consumption model of the Canadian residential sector. Each model was developed using the Stuttgart Neural Network Simulator (1998) and a systematic approach to identify the most suitable network architecture and set of variables to be included in the model. The approach included testing of various activation functions, various datasets scaled to different intervals and networks with different architectures.

The CDA Model also consists of three components. However, due to the characteristics and data requirements of the CDA method, these three components are configured differently than those of the NN Model. Each CDA component predicts the energy consumption of a household in one of the three predominantly used household fuels. Thus, the three components of the CDA Model are:

1. Electricity Model,
2. Natural gas Model,
3. Oil Model.

In the development process, all variables that influence the energy consumption for appliance, lighting and cooling, DHW heating, and space heating were included in the individual models. During the development process, the variables that were found to be statistically infeasible were eliminated, and the final set of variables that resulted in statistically robust models were identified. The regression analysis was conducted using the SYSTAT software (1998).

The prediction performance of the NN and CDA Models developed in this work were assessed by comparing their predictions to metered energy consumption data available for a subset of 2,749 households in the 1993 SHEU database. Since one of the important purposes of residential energy models is to study the characteristics of energy consumption, the capability of the two models to characterise energy consumption, as well as to study the impacts of socio-economic factors and energy saving measures on end-use energy consumption were assessed.

The NN and CDA Models developed in this work were also compared with an engineering model developed earlier by Farahbakhsh (1997) and Farahbakhsh *et al.* (1997, 1998) based on the 1993 SHEU database. The predictions of the NN and CDA Models were compared with those of the Engineering Model to assess the comparative accuracy, as well as the versatility of the three models.

The comparison of the predictions of the models indicated that all three models are capable of accurately predicting the energy consumption in the residential sector. The household and end-use energy consumption estimates of the three models were found to be close to each other, and also to the estimates reported by NRCan (2002) and OEE (2002). The accuracy of the predictions, calculated based on the average energy consumption in the subset of 2,749 SHEU households with

145

metered data is given in Table 7.1. These findings indicate that the NN Model has higher prediction performance than the CDA and the Engineering Models.

Table 7.1. Accuracy of the predictions of the NN, CDA, and Engineering Models in terms of fraction of variance (R^2) and coefficient of variation (CV)

	ALC		DHW Heating		Space Heating	
	R^2	CV	R^2	CV	R^2	CV
NN Model	0.909	2.094	0.871	3.337	0.908	1.871
CDA Model	0.795	3.343	0.814	4.052	0.892	2.007
Engineering Model	0.780	3.463	0.828	3.898	0.778	2.877

It was also found that the NN Model is capable of accurately predicting the energy consumption of individual households provided that the input units of these households are representative of the energy consumption of the households. The NN Model was found to be capable of accurately predicting the disaggregated energy consumption for central air conditioning, second refrigerator and boiler pump/furnace fan.

The NN Model is able to evaluate the effects of several socio-economic factors on end-use energy consumption. These include household income, dwelling type and ownership, number of children and adults, and size of area of residence. On the other hand, although theoretically possible, the CDA Model is unable to evaluate the effects of some of these socio-economic factors (dwelling ownership and size of area of residence), since the number of variables included in the CDA Model is limited due to statistical considerations. None of the socio-economic factors could be evaluated by the Engineering Model, since due to insufficient data on socio-economic factors in terms of occupancy and preference profiles in the 1993 SHEU database socio-economic variables were not included within the model structure of the existing Engineering Model. Thus from the perspective of assessing the impact of socio-economic factors, the NN Model is superior to both the CA and the Engineering Models.

In terms of estimating the impacts of the energy savings scenarios, the NN Model was found to be limited in its scope due to the limited number of variables that are included in the model. Since the NN Model includes the necessary variables, it can evaluate the following energy saving scenarios: (1) Insulating the hot water pipes, (2) Increasing the efficiency of the DHW heating systems, (3) Upgrading the glazing of the windows, (4) Increasing the efficiency of the SH equipment, (5) Lowering the overnight temperature. The accuracy of the predictions depends on the quality of information in the training dataset: as long as the households that undertake the energy saving measures are well represented in the training dataset, the accuracy is high (such as in the case of increasing the efficiency of the SH equipment). If however the households that undertake the energy saving measure are not well represented in the training dataset, the accuracy is low (such as in the case of upgrading the single glazed windows to double glazed). It should also be noted that since the NN Model predicts each end-use separately, it cannot evaluate the impact of an energy saving measure on other energy end-uses.

The Engineering Model, on the other hand, has a significantly higher level of flexibility in evaluating energy conservation measures, including the effects of energy efficiency measures on end-uses other than that is directly affected by the measure. This is due to the inherent advantage of the Engineering Model since it estimates the energy consumption using thermodynamics and heat transfer principles. The CDA Model however is not suitable to assess energy efficiency measures due to the limited number of variables the model can accommodate.

These results show that the NN method can be used for the following purposes:

- to estimate the end-use energy consumption in the residential sector,
- to categorize the household and end-use energy consumption to help the understanding of how energy is used in the residential sector,
- to evaluate the effects of (some) socio-economic factors on end-use energy consumption,
- to evaluate the impacts of (some) energy saving scenarios on end-use energy consumption.

The CDA Model, while simpler to apply than the NN Model and acceptably accurate in its overall predictions, is not flexible in evaluating end-uses, socio-economic factors and energy saving scenarios. Thus, the CDA Model has limited utility for modeling the energy consumption in the residential sector. The fundamental difference between the NN Model and the CDA Model is that the NN Model is less transparent to demonstrate the marginal effects of factors such as having or not having a particular appliance, DHW or space heating equipment. In comparison, the Engineering Model provides accurate estimates, has the highest level of flexibility in evaluating the impact of energy saving measures, but has difficulties with the inclusion of the socio-economic factors. The three models have their specific advantages and disadvantages. These can be summarized as shown in Table 7.2.

147

Table 7.2. Qualitative comparison of the models

	NN Model	CDA Model	Engineering Model
Prediction performance	High	Acceptable	Acceptable
Evaluation of Socio-economic factors	Easy	Limited	Difficult
Evaluation of energy saving scenarios	Limited	Poor	Excellent
Ease of use	Moderate	Easy	Requires extensive user expertise

In conclusion, the objectives set out in Section 3 of Chapter 1 were achieved as follows:

1. Two new end-use energy consumption models for the Canadian residential sector were developed using the 1993 SHEU database; one using the NN approach and the second using the CDA approach.

2. The annual average end-use and total household energy consumption values of the Canadian housing stock were estimated using the NN and the CDA Models, and the estimates were categorized based on dwelling type and age, province, and end-use fuel types.

3. The end-use and household energy consumption estimates of the NN and the CDA Models were compared with those of the Engineering Model and with metered data to assess the accuracy of the models.

4. The impacts of the energy saving scenarios were evaluated using the NN and CDA Models.

5. A comparative assessment of the NN, CDA, and Engineering Models was conducted.

2. Recommendations for Future Work

The recommendations for future work are as follows:

a) The NN Model was developed using the learning algorithms and the activation functions of the SNNS software. Other NN softwares with different learning algorithms and activation functions can be tested to increase the prediction performance of the model.

b) If a database with a sufficiently large number (*e.g.* > 5,000) of complete household data, including energy billing data for all types of fuels become available, energy consumption of all end-uses can be estimated using just one NN model. This would improve the capability to evaluate the impact of energy saving measures on all end-uses. Also, with such data a CDA Model can be developed to disaggregate the energy consumption of households with all fuel types in one model.

c) Various mathematical manipulations of the variables used in the CDA Model, such as taking the power or logarithm of variables, or cross multiplications of the variables, can be tested to reduce the multicollinearity and the non-constant variance problems faced in some components of the CDA Model.

d) Tobit regression analysis can handle discrete and continuous variables such as those in the datasets of the CDA Model. Thus, instead of the ordinary least square method it can be used to estimate the CDA Model.

e) The flexibility and scope of the models would improve if the size and the quality of the database of household information were improved. Therefore, a database representative of the national housing stock with detailed and accurate information on a large number of households (*e.g.* >10,000) would be needed if better models were to be developed.

f) The Engineering Model estimates the end-use energy consumption based on the thermodynamic and physical principles. Therefore, it is capable of evaluating the impact of a wide range of energy saving measures. However, it is difficult to incorporate the effects of the socio-economic factors into the Engineering Model, because detailed data are required to develop distribution functions for each socio-economic factor. On the other hand, the CDA and NN Models can reflect the effects of the socio-economic factors using the available data. Therefore, it may be possible to develop a hybrid model that uses the Engineering Model for physical and thermodynamic modeling and the CDA and NN Models for modeling of socio-economic factors.

References

Aigner, D.J., Sorooshian C., and Kerwin, P., (1984), "Conditional Demand Analysis for Estimating Residential End-Use Load Profiles", The Energy Journal, Vol 5, No. 3, pp. 81-97.

AlFuhaid, A.S., El-Sayed, M.A., Mahmoud, M.S., (1997), "Cascaded Artificial Neural Networks for Short-term Load Forecasting", IEEE Transactions on Power Systems, Vol. 12, No. 4, pp. 1524-1529.

Anstett, M., and Kreider, J.F., (1993), "Application of Neural Networking Models to Predict Energy Use", ASHRAE Transactions, Vol. 99, Part 1, pp. 505-517.

Aydinalp, M., Fung, A.S., and Ugursal, V.I., (1998), "Regression Analysis for Central Air Conditioning Data", Canadian Residential Energy End-Use Data and Analysis Centre, Technical Report, Prepared for NRCan, Halifax, N.S., Canada.

Aydinalp, M., Fung, A.S., and Ugursal, V.I., (2000), "Modeling of Residential Energy Consumption at the National Level", The Twelfth International Symposium on Transport Phenomena, July 16-20, Istanbul, Turkey.

Aydinalp, M., Ugursal, V.I., and Fung, A.S., (2001a), "Modeling of Residential Energy Consumption at the National Level", The International Conference on Industry, Engineering, and Management Systems (IEMS), March 5-7, Cocoa Beach, Florida, USA.

Aydinalp, M., Ugursal, V.I., and Fung, A.S., (2001b), "Predicting Residential Appliance, Lighting, and Space Cooling Energy Consumption Using Neural Networks", Fourth International Thermal Energy Congress, July 8-12, Izmir, Turkey

Aydinalp, M., Ugursal, V.I., and Fung, A.S., (2002a), "Modeling of the Appliance, Lighting, and Space Cooling Energy Consumption in the Residential Sector Using Neural Networks", Applied Energy, Vol. 71, Issue 2, pp. 87-110.

Aydinalp, M., Ugursal, V.I., and Fung, A.S., (2002b), "The Effects of the Socio-economic Factors on the Household Appliance, Lighting, and Cooling Energy Consumption", The Thirteenth International Symposium on Transport Phenomena, July 14-18, Victoria, B.C., Canada.

Bauwens, L., Fiebig, D., and Steel, M., (1994), "Estimating End-use Demand: A Bayesian Approach", Journal of Business and Economic Statistics, Vol. 12, No. 2, pp. 221 – 231.

Bierbaum, T.J., Case, M.A., Waston, P.A., Bush, G.L., and Pham, T., (2000), "Behaviour Recorder: Software to Record and Analyze Behaviour Data Using SAS and SYSTAT Statistical Software", Computers and Electronics in Agriculture, Vol. 29, No. 3, pp. 233 – 241.

Binfet, J., and Wilamowski, B.M., (2001), "Microprocessor implementation of fuzzy systems and neural networks", Proceedings of the International Joint Conference on Neural Networks IJCNN '01, Vol. 1, pp. 234 - 239.

Blaney, J.C., Inglis, M.R., and Janney, A.M., (1994), "Hourly Conditional Demand Analysis of Residential Electricity Use", 1994 ACEEE Summer Study on Energy Efficiency in Buildings, Panel 7, Pacific Grove, California, USA.

Brook, D., (1999), "Energy Note: Buying and Using Water Heaters-", Oregon State University, Extension Service, Corvallis, Oregon, USA. Available: http://www.energy.state.or.us/res/appntwht.pdf [2001, 14 November]

Caves, D.W., Herriges, J.A., Train, K.E., and Windle, R.J., (1987), "A Bayesian Approach to Combining Conditional Demand and Engineering Models of Electric Usage", Review of Economics and Statistics, Vol. 69, No. 3, pp. 438 – 448.

Chen, C.S., Tzeng, Y.M., and Hwang, J.C., (1996), "The Application of Artificial Neural Networks to Substation Load Forecasting", Electric Power Systems Research, Vol. 38, No. 2, pp. 153 – 160.

Chen, R.N., (1999), "The Cumulative q Interval Curve as a Starting Point in Disease Cluster Investigation", Statistics in Medicine, Vol. 18, No. 23, pp. 3299-3307.

Cohen, D.A., and Krarti, M., (1995), "A Neural Network Modeling Approach Applied to Energy Conservation Retrofits", Proceedings of the Building Simulation Fourth International Conference, pp. 423 - 430.

Conachen, J., (2001), "Existing Single-Pane Wood Windows Become Energy Efficient", Inventions and Innovations Program, US Department of Energy. Available: www.oit.doe.gov/inventions/pdfs/portfolio/other/buildings/biglass.pdf [2002, 12 February]

Curtiss, P.S., Shavit, G., and Kreider, J.F., (1996), "Neural Networks Applied to Buildings- A Tutorial and Case Studies in Prediction and Adaptive Control", ASHRAE Transactions, Vol. 102, No. 1, pp. 1141-1146.

Del Frate, F., and Schiavon, G., (1995), "Retrieval of Atmospheric Parameters from Radiometric Measurements Using Neural Networks", International Geoscience and Remote Sensing Symposium, IGARSS '95, 'Quantitative Remote Sensing for Science and Applications', Vol. 2, pp. 1134.

Dodier, R., and Henze, G., (1996), "Statistical Analysis of Neural Network as Applied to Building Energy Prediction", Proceedings of the ASME ISEC, San Antonio, TX, USA, pp. 495 - 506.

El-Fergany, A.A., Yousef, M.T., and El-Alaily, A.A., (2001), "Fault Diagnosis in Power Systems - Substation Level- Through Hybrid Artificial Neural Networks and Expert System", Proceedings of the IEEE Power Engineering Society Transmission and Distribution Conference and Exposition, IEEE/PES 2001, Vol. 1, pp. 207-211.

Energy Efficiency and Renewable Energy Clearinghouse (EREC), (1997), "Automatic and Programmable Thermostats", US Department of Energy, National Renewable Energy Laboratory, Merrifield, VA, USA. Available: http://www.eren.doe.gov/erec/factsheets/thermo.pdf [2002, 27 February]

Environment Canada, (1999), "Canadian Climate and Water Information", Canadian Meteorological Centre, Ottawa, Ontario, Canada. Available: http://www.msc-smc.ec.gc.ca/climate/index_e.cfm [1999, 29 June]

Environment Canada, (2001), "Greenhouse Gas Emissions", Greenhouse Gas Division, Environment Canada, Ottawa, Ontario, Canada. Available: http://www.ec.gc.ca/pdb/ghg/ghg_home_e.cfm [2002, 10 May]

EPRI, (1989), "Residential End-use Energy Consumption: A Survey of Conditional Demand Analysis", Report No. CU-6487, Palo Alto, CA, USA.

Farahbakhsh, H., (1997), "Residential Energy and Carbon Dioxide Emissions Model for Canada", Master Thesis, Technical University of Nova Scotia, Department of Mechanical Engineering, Halifax, N.S., Canada.

Farahbakhsh, H., Fung, A.S., and Ugursal, V.I., (1997), "Space Heating Thermal Requirements and Unit Energy Consumption of Canadian Homes in 1993", Canadian Residential Energy End-Use Data and Analysis Centre, Final Report, Prepared for NRCan, Halifax, N.S., Canada.

Farahbakhsh, H., Ugursal, V.I., and Fung, A.S., (1998), "A Residential End-use Energy Consumption Model for Canada", International Journal of Energy Research, Vol. 22, No.13, pp.1133 -1143.

Fausett, L., (1994), "Fundamentals of Neural Networks", Prentice Hall, Englewoods Cliffs, N.J., USA.

Feuston, B.P., and Thurtell, J.H., (1994), "Generalized Nonlinear Regression with Ensemble of Neural Nets: The Great Energy Predictor Shootout", ASHRAE Transactions, Vol. 100, No. 2, pp. 1075-1080.

Fiebig, D.G., Bartels R., and Aigner D.J., (1991), "A Random Coefficient Approach to the Estimation of Residential End-Use Load Profiles", Journal of Econometrics, Vol. 50, pp. 297 – 327.

Fung, A.S., Farahbakhsh, H., and Ugursal, V.I., (1997), "Unit Energy Consumption of Major Household Appliances in Canada", Canadian Residential Energy End-Use Data and Analysis Centre, Technical Report, Prepared for NRCan, Halifax, N.S., Canada.

Gemperline, P.J., (2000), "Powerful Statistical Analysis with SYSTAT 9.0", Analytical Chemistry, Vol. 72, No. 9, pp. 362A-362A.

Gottsche, F.M., and Olesen, F.S., (2001), "Evolution of Neural Networks for Radiative Transfer Calculations in the Terrestrial Infrared", Remote Sensing of Environment, Vol. 80, No. 1, pp. 157 – 164.

Guler, B., Fung, A.S., Aydinalp, M., and Ugursal, V.I., (2001), "Impact of Energy Efficiency Upgrade Retrofits on the Residential Energy Consumption in Canada", International Journal of Energy Research, Vol. 25, pp. 785 – 792.

Hairston, J.E., (1995), "Water Quality and Pollution Control Handbook", Section 1.3: Conserving Water: Installing Water-Saving Devices, Alabama Cooperative Extension System, Available: http://www.aces.edu/department/extcomm/publications/anr/anr-790/WQ1.3.2.pdf [2001, 14 November]

Hassoun, M.H., (1995), "Fundamentals of Artificial Neural Networks", MIT Press, Cambridge, Massachusetts, USA.

Highley, D.D., and Hilmes, T., (1993), "Load Forecasting by ANN", IEEE Computer Applications in Power, pp. 10 – 15.

Hsiao, C., Mountain, D.C., and Illman, K.H., (1995), "Bayesian Integration of End-Use Metering and Conditional Demand Analysis", Journal of Business and Economic Statistics, Vol 13, No. 3, pp. 315-326.

Johnston, J., and DiNardo, J., (1997), "Econometric Methods", McGraw-Hill. New York, N.Y., USA.

Kawashima, M., (1994), "Artificial Neural Network Backpropagation Model with Three-Phase Annealing Developed for the Building Energy Predictor Shootout", ASHRAE Transactions, Vol. 100, Part. 2, pp. 1095 - 1103.

Kellas, C., (1993), "Conditional Demand Analysis in Manitoba 1993", Canadian Electrical Association Conference, May 1993, Halifax, Nova Scotia, Canada.

Kiartzis, S.J., Bakirtzis, A.G., and Petridis, V., (1995), "Short-term Forecasting Using NNs", Electric Power Systems Research, Vol. 33, pp. 1-6.

Kornbrot, D., (1999), "Statistical Software for Microcomputers: SYSTAT 8.0", British Journal of Mathematical and Statistical Psychology, Vol. 52, pp. 143-145.

Krarti, M., Kreider, J.F., Cohen, D., and Curtiss, P., (1998), "Estimation of Energy Saving for Building Retrofits Using Neural Networks", Journal of Solar Energy Engineering, Vol. 120, pp. 211-216.

Kreider, J.F., and Wang, X.A., (1991), "Artificial Neural Networks Demonstrations for Automated Generation of Energy Use Predictors for Commercial Buildings", ASHRAE Transactions, Vol. 97, Part. 1, pp. 775-779.

Kreider, J.F., and Wang, X.A., (1992), "Improved Artificial Neural Networks for Commercial Building Energy Use Prediction", Solar Engineering, ASM, Vol. 1, pp. 361 - 366.

Kreider, J.F., and Haberl. J.S., (1994), "Predicting Hourly Building Energy Use: The Great Energy Predictor Shootout- Overview and Discussion of Results", ASHRAE Transactions, Vol. 100, Part. 2, pp. 1104-1118.

Kreider J.F., Claridge, D.E., Curtiss, P., Dodier, R., Haberl, J.S., and Krarti, M., (1995), "Building Energy Use Prediction and System Identification Using Recurrent Neural Networks", Journal of Solar Energy Engineering, Vol. 117, pp. 161 – 166.

Lafrance, G., and Perron, D., (1994), "Evolution of Residential Electricity Demand by End-Use in Quebec 1979-1989: A Conditional Demand Analysis", Energy Studies Review, Vol. 6, No. 2, pp. 164-173.

Leach, C., Freshwater, K., Aldridge, J., and Sunderland, J., (2001), "Analysis of Repertory Grids in Clinical Practice", British Journal of Clinical Psychology, Vol. 40, Part 3, pp. 225 – 248.

McCulloh, W.S. and Pitts, W.H., (1943), "A Logical Calculus of the Ideas Immanent in Nervous Activity", Bulletin of Mathematical Biophysics, Vol. 5, pp. 115-33.

Miller, W.C., Swensen, T., and Wallace, J.P., (1998), "Derivation of Prediction Equations for RV in Overweight Men and Women", Medicine and Science in Sports and Exercise, Vol. 30, No. 2, pp. 322-327.

Natural Resources Canada (NRCan), (1994), "200-House Audit Project", Ottawa, Ontario, Canada.

Natural Resources Canada (NRCan), (1995), "EnerGuide Appliance Directory-1994", Energy Publications, Ottawa, Ontario, Canada.

Natural Resources Canada, (NRCan), (1996), "HOT2000 Batch V7.13 User Manual", Ottawa, Ontario, Canada.

Natural Resources Canada, (NRCan), (2002), "EnerGuide for Houses Rating and Label", Natural Resources Canada, Ottawa, Ontario, Canada. Available: http://oee.nrcan.gc.ca/houses-maisons/english/e51.cfm#soc [2002, 12 July]

Office of Energy Efficiency (OEE), (2002), "Energy Efficiency Trends- Summary Tables", Natural Resources Canada, Ottawa, Ontario, Canada. Available: http://oee1.nrcan.gc.ca/neud/dpa/summary_tables.cfm [2002, 14 June]

Okon K., Tomaszewska, R., and Stachura, J., (2001), "Application of Neural Networks to the Classification of Pancreatic Intraductal Proliferative Lesions", Analytical Cellular Pathology, Vol. 23, No. 3-4, pp. 129 – 136.

Olofsson, T.; and Andersson, S., (2001), "Long-term Energy Demand Predictions Based on Short-term Measured Data", Energy and Buildings, Vol. 33, No. 2, pp. 85 – 91.

Park, D.C., El-Sharkawi, M.A., Marks, R.J., Atlas, L.E., and Damborg, M.J., (1991), "Electric Load Forecasting Using an ANN", IEEE Transactions on Power Systems, Vol. 6, No. 2, pp. 442-449.

Parti, M., and Parti C., (1980), "The Total and Appliance-specific Conditional Demand for Electricity in the Household Sector", Bell Journal of Economics, Vol. 11, pp. 309-321.

Peng, T.M., Hubele, N.F. and Karady, G.G., (1992), "Advancement in the Application of NN for Short-term Load Forecasting", IEEE Transactions on Power Systems, Vol. 7, No. 1, pp. 250-257.

Pontikakis, N., and Ruth, D.W., (1999), "Modeling of Residential Hot Water Systems", Transactions of CSME, Vol. 23, No. 1B, pp. 197-212.

Riedmiller, M., and Braun, H., (1993), "A Direct Adaptive Method for Faster Backpropagation Learning: The RPROP Algorithm", The Proceedings of the IEEE International Conference on Neural Networks, Vol. 1, pp. 586-591.

Rumelhart, D.E., and McClelland, J.L., (1986), "Parallel Distributed Processing", The MIT Press, Cambridge, Massachusetts, USA.

Scanada Consultants Limited, (1992), "Statistically Representative Housing Stock", Final Report, Prepared for Canada Mortgage and Housing Corporation, Ottawa, Ontario, Canada.

Sindt, M., Stephan, B., Schneider, M., and Mieloszynski, J.L., (2001), "Chemical Shift Prediction of P-31-NMR Shifts for Dialkyl or Diaryl Phosphonates", Phosphorous Sulphur and Silicon and the Related Elements, Vol. 174, pp. 163-175.

SNNS, (1998), "User Manual", Version 4.2, University of Stuttgart, Stuttgart, Germany. Available: http://www-ra.informatik.uni-tuebingen.de/SNNS/ [1999, 12 April].

Statistics Canada, (1993), "Microdata User's Guide", The Survey of Household Energy Use, Ottawa, Ontario, Canada.

Stevenson, J.S., (1994), "Using Artificial Neural Nets to Predict Building Energy Parameters", ASHRAE Transactions, Vol. 100, No.2, pp.1081-1087

SYSTAT, (1998), "SYSTAT- Statistics", SPSS Inc., Chicago, Illinois, USA.

Swamy, P.A., (1970), "Efficient Inference in a Random Coefficient Regression Model", Econometrica, Vol. 38, No. 2, pp. 311 – 323.

Tomlinson, J.J., and Rizy, D.T., (1998), "Bern Clothes Washer Study", Energy Division Oak Ridge National Laboratory, Final Report, Prepared for US Department of Energy, Oak Ridge, Tennessee, USA. Available: http://www.eren.doe.gov/buildings/emergingtech/pdfs/bernrpt.pdf [2001, 5 November]

Train, K.E., (1992), "An Assessment of the Accuracy of Statistically Adjusted Engineering (SAE) Models of End-use Load Curves", Energy Journal, Vol. 17, No. 7, pp. 713-723.

Ugursal, V.I., and Fung, A.S., (1994), "Energy Efficiency Technology Impact – Appliances, Volume 1, Report Prepared for Canada Mortgage and Housing Corporation, Ottawa, Ontario, Canada.

Ugursal, V.I., and Fung, A.S., (1996), "Impact of Appliance Efficiency and Fuel Substitution on Residential End-use Energy Consumption in Canada", Energy and Buildings, Vol. 24, No.2, pp. 137-146.

Ugursal, V.I., and Fung, A.S., (1998), "Residential Carbon Dioxide Emissions in Canada: Impact of Efficiency Improvements and Fuel Substitution", Global Environmental Change, Vol. 8, No. 3, pp. 263-273.

Weisberg, S., (1985), "Applied Linear Regression", Second Edition, John Wiley and Sons, New York, N.Y., USA.

Wittmann, C., Schmid, R.D., Loffler, S., and Zell, A., (1997), "Application of a Neural Network for Pattern Recognition of Pesticides in Water Samples by Different Immunochemical Techniques", Immunochemical Technology for Environmental Applications ACS Symposium Series, pp. 343-360.

Zeller, A., (1971), "An Introduction to Bayesian Inference in Econometrics", John Wiley&Sons, Inc., New York, N.Y., USA.

APPENDICES

APPENDIX A

NN LEARNING ALGORITHMS

1. BACKPROPAGATION ALGORITHMS

The error signal of an output layer neuron is defined by (Rumelhart and McClelland, 1986; Fausett, 1994):

$$e_k = t_k - y_k \qquad \text{(A.1)}$$

where,

t_k : target value

y_k : predicted value

The total error is the sum of the square of the error signals for all of the output units:

$$E = \frac{1}{2}\sum_{k=1}^{m} e_k^{\,2} \qquad \text{(A.2)}$$

where,

E : total error or error function

m : number of output units

The factor of 1/2 in Equation A.2 is used for convenience in calculating derivatives later. Since an arbitrary constant will appear in the final result, the presence of this factor does not invalidate the derivation (Freeman and Skapura, 1991).

The input of an output layer neuron expressed as:

$$net_k = \sum_{j=1}^{p} z_j w_{jk} + b_k \qquad \text{(A.3)}$$

where,

net_k : total input of the output layer neuron k

z_j : input to the output layer neuron k from hidden layer neuron j

w_{jk} : weight between the hidden layer neuron j and output layer neuron k

b_k : bias value of the output layer neuron k

p : number of neurons in the hidden layer

Hence, the output of an output layer neuron becomes:

$$y_k = f(net_k) \qquad \text{(A.4)}$$

The weights between the hidden layer neurons and output layer neurons are corrected by applying a weight correction term, Δw_{jk}, which is proportional to the partial derivative $\partial E/\partial w_{jk}$. By using the chain rule,

159

$$\frac{\partial E}{\partial w_{jk}} = \frac{\partial E}{\partial e_k} \frac{\partial e_k}{\partial y_k} \frac{\partial y_k}{\partial net_k} \frac{\partial net_k}{\partial w_{jk}} \tag{A.5}$$

Differentiating both sides of Equation A.2 with respect to e_k:

$$\frac{\partial E}{\partial e_k} = e_k \tag{A.6}$$

Differentiating both sides of Equation A.1 with respect to y_k:

$$\frac{\partial e_k}{\partial y_k} = -1 \tag{A.7}$$

Differentiating both sides of Equation A.4 with respect to net_k:

$$\frac{\partial y_k}{\partial net_k} = f'(net_k) \tag{A.8}$$

Finally, differentiating both sides of Equation A.3 with respect to w_{jk}:

$$\frac{\partial net_k}{\partial w_{jk}} = z_j \tag{A.9}$$

The substitution Equations A.6 to A.9 into Equation A.5 yields:

$$\frac{\partial E}{\partial w_{jk}} = -e_k f'(net_k) z_j \tag{A.10}$$

A new term called the error information term, δ_k, is defined as:

$$\delta_k = -\frac{\partial E}{\partial net_k} \tag{A.11}$$

By using Equations A.6 to A.8 and chain rule, δ_k is expressed as:

$$\delta_k = -\frac{\partial E}{\partial e_k} \frac{\partial e_k}{\partial y_k} \frac{\partial y_k}{\partial net_k} \tag{A.12}$$

$$= e_k f'(net_k)$$

$$= (t_k - y_k) f'(net_k)$$

The calculation of error information terms for the hidden layer neurons is complicated, since there is no specified desired response for the hidden layer neurons. Even though hidden layer neurons are not directly accessible, they share responsibility for any error made at the output of the network. Thus, the error signal for a hidden layer neuron is determined in terms of the error signals of all the neurons to which the hidden layer neuron is directly connected. Hence, the error information term can be expressed as:

$$\delta_j = -\frac{\partial E}{\partial net_j} \tag{A.13}$$

$$= -\frac{\partial E}{\partial z_j} \frac{\partial z_j}{\partial net_j}$$

$$= -\frac{\partial E}{\partial z_j} f'(net_j)$$

Differentiating Equation A.2 with respect to the functional signal $f(net_j) = z_j$:

$$\frac{\partial E}{\partial z_j} = \sum_{k=1}^{m} e_k \frac{\partial e_k}{\partial z_j} \tag{A.14}$$

By using the chain rule for the partial derivative $\partial e_k/\partial z_j$, Equation A.14 is rewritten as:

$$\frac{\partial E}{\partial z_j} = \sum_{k=1}^{m} e_k \frac{\partial e_k}{\partial net_k} \frac{\partial net_k}{\partial z_j} \tag{A.15}$$

The substitution of Equation A.4 into Equation A.1 yields:

$$e_k = t_k - y_k \tag{A.16}$$

$$= t_k - f(net_k)$$

Hence the partial derivative $\partial e_k/\partial net_k$ becomes:

$$\frac{\partial e_k}{\partial net_k} = -f'(net_k) \tag{A.17}$$

Also, the partial derivative $\partial net_k/\partial z_j$ can be written by using Equation A.3 as:

$$\frac{\partial net_k}{\partial z_j} = w_{jk} \tag{A.18}$$

The substitution of Equations A.17 and A.18 into Equation A.15 yields:

$$\frac{\partial E}{\partial z_j} = -\sum_{k=1}^{m} e_k f'(net_k) w_{jk} \tag{A.19}$$

$$= -\sum_{k=1}^{m} (t_k - y_k) f'(net_k) w_{jk}$$

161

$$= -\sum_{k=1}^{m} \delta_k w_{jk}$$

Finally, the substitution of Equation A.19 into Equation A.13 yields:

$$\delta_j = f'(net_j) \sum_{k=1}^{m} \delta_k w_{jk} \tag{A.20}$$

(a) Standard (Plain Vanilla) Backpropagation

For each output neuron, the error information term, δ_k, is computed, and the weights are adjusted as:

$$\Delta w_{jk} = -\eta \frac{\partial E}{\partial w_{jk}} \tag{A.21}$$

$$= \eta (t_k - y_k) f'(net_k) z_j$$

$$= \eta \, \delta_k \, z_j$$

where,

η :   learning parameter

Δw_{jk} :   weight correction term for weights between the hidden units and output units

The learning parameter specifies the step width of the gradient descent, usually a small number in the order of 0.05 to 0.25 (SNNS, 1998). The use of the minus sign in Equation A.21 accounts for gradient descent in weight space, i.e., seeking a direction for weight change that reduces the value of E.

For the weights to the hidden units:

$$\Delta v_{ij} = \eta \delta_j x_i \tag{A.22}$$

$$= \eta f'(net_j) x_i \sum_{k=1}^{k} \delta_k w_{jk}$$

where,

Δv_{ij} : weight correction term for weights between the input units and hidden units

(b) Enhanced Backpropagation

The weight correction term is computed by using a momentum term and a flat spot elimination value for the weights to the output units as:

$$\Delta w_{jk}(t+1) = \eta \, \delta_k \, z_j + \mu \, \Delta w_{ik}(t) \qquad (A.23)$$

For the weights to the hidden units:

$$\Delta v_{ij}(t+1) = \eta \, \delta_j \, x_i + \mu \, \Delta v_{ij}(t) \qquad (A.24)$$

where,

μ : constant specifying the influence of the momentum.

The typical values for the momentum term, μ, are between zero and one (SNNS, 1998). The flat spot elimination value, c, is added to the derivative of the activation function. For the weights to the output units:

$$\delta_k = f'(net_k + c) \, (t_k - y_k) \qquad (A.25)$$

For the weight to the hidden units:

$$\delta_j = f'(net_j + c) \, (t_k - y_k) \qquad (A.26)$$

where,

c : flat spot elimination value with a typical value of 0.1 (SNNS, 1998).

2. QUICKPROP

The partial derivatives of the error function with respect to the given weight is summed over all training patterns, and is calculated as:

$$S(t) = \sum_{p=1}^{p} \frac{\partial E_p}{\partial w_{jk}} \qquad (A.29)$$

where,

S(t) : partial derivative of the error function

p : number of training patterns.

For a weight from a hidden unit to an output unit:

163

$$S_{jk}(t) = -\sum_{p=1}^{p} \delta_{k(p)} z_{j(p)} \tag{A.30}$$

and similarly, for a weight from an input unit to a hidden unit:

$$S_{ij}(t) = -\sum_{p=1}^{p} \delta_{j(p)} x_{i(p)} \tag{A.31}$$

The initial weight change is calculated by standard (vanilla) backpropagation as:

$$\Delta w(0) = -\eta S(0) \tag{A.32}$$

The new weight change for weights to the output units is defined to be:

$$\Delta w_{jk}(t) = \Delta w_{jk}(t-1) \times \frac{S_{jk}(t)}{S_{jk}(t-1) - S_{jk}(t)} \tag{A.33}$$

where,

$\Delta w_{jk}(t)$: weight correction term for the weights to the output units

$\Delta w_{jk}(t-1)$: previous weight correction term for the weights to the output units

$S_{jk}(t)$: partial derivative of the error function by w_{jk}

$S_{jk}(t-1)$: previous partial derivative of the error function by w_{jk}

The new weight change for weights to the hidden units is defined to be:

$$\Delta v_{ij}(t) = \Delta v_{ij}(t-1) \times \frac{S_{ij}(t)}{S_{ij}(t-1) - S_{ij}(t)} \tag{A.34}$$

where,

$\Delta v_{ij}(t)$: weight update value for the weights to the hidden units

$\Delta v_{ij}(t-1)$: previous weight correction term for the weights to the hidden units

$S_{ij}(t)$: partial derivative of the error function by v_{ij}

$S_{ij}(t-1)$: previous partial derivative of the error function by v_{ij}

The weight correction term, Δ, is used to determine the coefficients of the polynomial. At the next step of iteration, the weight parameter is moved to the minimum of the parabola.

If the current slope term is in the same direction as the previous slope, and it has the same size

164

or larger in magnitude than the previous slope, then the weight change would be infinite, or the weight would be moved away from the minimum and toward a maximum of the error (Fausett, 1994). To prevent this, weight change is limited by multiplying the previous step by the maximum growth parameter, ρ. The typical values of maximum growth parameter, ρ, are between 0.75 and 2.25 (SNNS, 1998).

A further refinement is used when the current slope has the same sign as the previous one. In this case, the current slope is multiplied by the weight decay term, v, to prevent the weight changed from being frozen (Fausett, 1994). The typical value of weight decay term, v, is 0.0001 (SNNS, 1998).

3. RESILIENT PROPAGATION (RPROP)

The update value, ϕ_{jk}, for each weight is computed as:

$$
\phi_{jk}^{(t)} = \begin{cases}
\alpha^+ \, \phi_{jk}^{(t-1)}, & \text{if } \dfrac{\partial E}{\partial w_{jk}}^{(t-1)} \dfrac{\partial E}{\partial w_{jk}}^{(t)} > 0 \\[3mm]
\alpha^- \, \phi_{jk}^{(t-1)}, & \text{if } \dfrac{\partial E}{\partial w_{jk}}^{(t-1)} \dfrac{\partial E}{\partial w_{jk}}^{(t)} < 0 \\[3mm]
\phi_{jk}^{(t-1)}, & \text{else}
\end{cases}
\tag{A.35}
$$

where,

$\phi_{jk}^{(t)}$: update value

$\phi_{jk}^{(t-1)}$: previous update value

$\dfrac{\partial E}{\partial w_{jk}}^{(t-1)}$: previous partial derivative of the error function of the corresponding weight w_{jk}

summed over all training patterns

$\dfrac{\partial E}{\partial w_{jk}}^{(t)}$: partial derivative of the error function of the corresponding weight w_{jk} summed

over all training patterns

α : increase/decrease factor with a value of $0 < \alpha^- < 1 < \alpha^+$.

After the update value of each weight is computed, the weight correction term is calculated as:

$$\Delta w_{jk}^{(t)} = \begin{cases} -\phi_{jk}^{(t)}, & \text{if } \dfrac{\partial E}{\partial w_{jk}}^{(t)} > 0 \\[2ex] +\phi_{jk}^{(t)}, & \text{if } \dfrac{\partial E}{\partial w_{jk}}^{(t)} < 0 \\[2ex] 0, & \text{else} \end{cases} \qquad (A.36)$$

The increase and decrease factors are set to fixed values, i.e. $\alpha^- = 0.5$ and $\alpha^+ = 1.2$, in order to reduce the number of adjustable parameters (Riedmiller and Braun, 1993; SNNS, 1998).

Weight decay term, β, is introduced to the error function to reduce the size of the weight:

$$E = \sum_{k=1}^{m}(t_k - y_k)^2 + 10^{-\beta}\sum_{k=1}^{m} w_{jk}^2 \qquad (A.37)$$

When learning starts, all update values are set to an initial value, $\phi_{initial}$, which has a typical value of 0.01 (SNNS, 1998). In order to prevent the weights from becoming too large, the maximum weight step determined by the size of the update value is limited. The default value for the maximum update value, ϕ_{max}, is 50 (SNNS, 1998).

166

APPENDIX B

ANALYSIS OF ENERGY BILLING DATA

ENERGY BILLING DATA

With the permission of the occupants, Statistics Canada obtained the complete year energy billing data of a subset of 3,341 households in the 1993 SHEU database from their fuel suppliers and utility companies. There are a total of 5,048 bills in the form of 3,400 electricity, 1,397 natural gas, and 251 oil. However, not all of the billing data appear to be reliable as explained below. An analysis of the data indicated that out of these 5,048 bills, a total of 3,298 bills from 2,749 households can be considered reliable. Within these 2,749, there are 2,050 households with electricity bills, 1,012 households with natural gas bills, and 236 households with oil bills. The number of households with electricity, oil, and natural gas bills is given in Table B.1.

Table B.1. Number of households with electricity, oil, and natural gas bills

Fuel Type of the Energy Billing Data	Number of Households
Electricity	1,502
Natural gas	540
Oil	158
Electricity and natural gas	471
Electricity and oil	77
Natural gas and oil	1

1. Electricity:

- Monthly: Out of 2,398, there are 724 bills with missing data. Thus, the number of bills reduces to 1,674.
- Bi-monthly: Out of 734, there are 138 bills with missing data. Thus, the number of bills reduces to 596.
- Annual: There are 238 bills.
- None: There are 29 bills with no information.

The total number of electricity bills becomes 2,508 (1674 + 596 + 238 = 2,508). Considering the fact that a typical household usually has at least a refrigerator and lighting, a minimum electricity load comprising average refrigerator and lighting load was assumed to be the criterion for discarding the unacceptable billing data. Knowing that on average a refrigerator consumes about 1,330 kWh/yr, and an average household has a lighting consumption of about 1,770 kWh/yr (Fung et al., 1997; Ugursal and Fung, 1994), an annual energy consumption of 3,000 kWh/yr was assumed as the minimum electrical consumption of a household. After excluding these bills from the analysis, the total number of bills reduces to 2,341.

168

Since only single detached and attached dwellings in the 1993 SHEU database is used in this work, the billing data for apartments and mobile homes are discarded. Thus, the number of bills is reduced to 2,050.

The electricity consumption in a household consists of space heating, DHW heating, space cooling, appliance and lighting consumption. The number of households which use each of the end-uses is given in Table B.2. Thus, there are only 633 households that use electricity only for lighting and appliances, *i.e.*, these households do not have cooling and do not use electricity for space or DHW heating.

Table B.2. Electricity bill distribution based on end-use consumption

End-use Consumption	Number of Households
Appliance and lighting	633
ALC	355
ALC and space heating	25
ALC and DHW heating	506
ALC, space, and DHW heating	531

The bi-monthly bills are converted to monthly bills by dividing the bi-monthly consumption into two and combined with the monthly bills. When bills from mobile homes and apartments are excluded, the number of monthly bills is reduced to 1,891. The distribution of households with monthly electricity bills for each end-use is given in Table B.3. Thus, there are only 523 households with monthly electricity bills that use electricity only for lighting and appliances, *i.e.*, these households do not have cooling and do not use electricity for space or DHW heating.

Table B.3. Monthly electricity bill distribution based on end-use consumption

End-use Consumption	Number of Households
Appliance and lighting	523
ALC	338
ALC and space heating	21
ALC and DHW heating	484
ALC, space, and DHW heating	525

(a) Natural Gas:

- Monthly: Out of 1,113, there are 321 bills with missing data. Thus, the number of bills reduces to 792.
- Bi-monthly: There are 20 bills.
- Annual: There are 217 bills.
- None: There are 47 bills with no information.

The total number of natural gas bills becomes 1,029 (792 + 20 + 217 = 1,029). Since only single detached and attached dwellings in the 1993 SHEU database are used in this work, the billing data for apartments and mobile homes are discarded. Thus, the number of bills is reduced to 1,012.

The natural gas consumption in a household mainly consists of space and DHW heating. Natural gas is also used for ranges, clothes dryers, fireplaces, backup furnaces for heat pumps, supplementary heaters, and pool heaters; but the saturation of the households using natural gas for these end-uses is very low. Hence, in this work natural gas consumption for space and DHW heating are considered. There are seven households with natural gas bills but do not have natural gas space or DHW heating, thus the number of households with natural gas bills reduces to 1,005. The distribution of households for each end-use are given in Table B.4. Thus, there are 63 households that use natural gas for only space heating, 27 households using natural gas for only DHW heating, and 915 households using natural gas for both space and DHW heating.

Table B.4. Natural gas bill distribution based on end-use consumption

End-use Consumption	Number of Households
Space heating	63
DHW heating	27
DHW and space heating	915

The bi-monthly bills are converted to monthly bills by dividing the bi-monthly consumption into two and combined with the monthly bills. When bills from mobile homes and apartments are excluded, the number of monthly bills is reduced to 791. The distribution of households with monthly natural gas bills for each end-use is given in Table B.5. Thus, there are 35 households with monthly bills that use natural gas for only space heating, 11 households using natural gas for only DHW heating, and 745 households using natural gas for both space and DHW heating.

170

Table B.5. Monthly natural gas bill distribution based on end-use consumption

End-use Consumption	Number of Households
Space heating	35
DHW heating	11
DHW and space heating	745

(b) Oil:

- Separate deliveries full season: There are 95 bills.
- Total only: There are 152 bills.
- None: There are 4 bills with no information.

The total number of oil bills becomes 247 (95 +152 = 247). Since only single detached and attached dwellings in the 1993 SHEU database are used in this work, the billing data for apartments and mobile homes are discarded. Thus, the number of bills is reduced to 236.

The oil consumption in a household mainly consists of space and DHW heating. Oil is also used for ranges, backup furnaces for heat pumps, supplementary heaters, and pool heaters; but the saturation of the households using oil for these end-uses is very low. Hence, in this work oil consumption for space and DHW heating are considered. There are six households with oil bills but do not use oil as the fuel for space or DHW heating, thus the number of households with oil bills reduces to 230. The distribution of households for each end-use are given in Table B.6. Thus, there are 90 households that use oil for only space heating, three households using oil for only DHW heating, and 137 households using oil for both space and DHW heating.

Table B.6. Oil bill distribution based on end-use consumption

End-use Consumption	Number of Households
Space heating	90
DHW heating	3
DHW and space heating	137

171

APPENDIX C

DEVELOPMENT OF THE DHW NETWORK DATASET

DHW NETWORK DATASET

Billing data obtained from fuel suppliers and utility companies were available for a subset of 3,341 households in the 1993 SHEU database. As shown in Appendix B, a total of 3,298 bills from 2,749 households were considered reliable. Out of these 2,749 households, there are 2,050 households with electricity bills, 1,012 households with natural gas bills, and 236 households with oil bills. The screening processes used in the selection of the households with the billing data for the development of the DHW network dataset are presented in the following sections.

1. Households with Electricity Bills

There are 2,050 households with electricity bills. The distribution of these households based on their end-use electricity consumption is given in Table B.2 of Appendix B. As seen in Table B.2 of Appendix B, there are 506 households using electricity for ALC and DHW heating, and 531 households using electricity for ALC, DHW, and SH.

The electricity consumption data for the 506 households using electricity for ALC and DHW heating were screened as follows:

1. The ALC electricity consumption estimated by the ALC NN model was deducted from the annual electricity consumption of the 506 households, and 94 households were found to have negative values. This shows that these households consumed less energy for ALC and DHW heating than the estimated values. This could be due to the vacancy of the dwellings for long periods of time or an error associated with the billing data. Thus, these 94 households were excluded from the analysis.

2. The minimum likely annual electricity consumption, Q_{min} (MJ/yr), of a household for DHW heating was estimated using Equation C.1 assuming that:

   - the household has one occupant,
   - DHW heating system efficiency is 0.96,
   - average annual ground temperature is 12 $^\circ$C (285K).

$$Q_{min} = \frac{mc\Delta T}{EF} \qquad (C.1)$$

where m : annual DHW consumption, kg/yr

c : $4.184 \ 10^{-3}$ MJ/kgK

ΔT: [318 K – Average annual ground temperature, K][13]

EF: efficiency of the DHW heating system

The annual DHW consumption (m) of a household can be estimated from Equation C.2 (NRCan, 1996).

$$m \text{ (kg/d)} = [85 \text{ l/d} + (35 \text{ l/d-people} * \text{no. people})] * 365 \text{ d/yr} * 0.997 \text{ kg/l} \qquad \text{(C.2)}$$

Using Equations C.1 and C.2, the minimum likely annual electricity consumption of a household for DHW heating was calculated as 6,280.7 MJ/yr $\approx$ 1,745 kWh/yr. There were 71 households with DHW heating consumption values less than 1,745 kWh. These 71 households were therefore excluded from the analysis.

Thus, the number of households with electricity bills and using electricity for ALC and DHW heating was reduced to 341 after excluding the 165 (= 94 + 71) households as shown above.

There are 531 households using electricity for ALC, and DHW and SH in the billing dataset. The ALC electricity consumption of these households was estimated by the ALC NN model. To include these households in the dataset used for the development of the DHW network, the amount of electricity consumed for SH by these households should be estimated.

Assuming that the monthly ALC electricity consumption of these households was constant throughout the year, the annual ALC electricity consumption estimated by the ALC NN model for these households was divided by twelve and deducted from the monthly electricity consumption. Therefore, the remaining monthly electricity consumption of these households accounted for DHW and SH electricity consumption.

It was assumed that during summer months (*i.e.* June, July, and August), the electricity consumption of these households accounted only for DHW heating, and there was not any electricity consumption for SH. The minimum monthly electricity consumption in these summer months can be taken as the household's monthly DHW heating electricity consumption. Therefore, the annual DHW heating electricity consumption can be calculated by multiplying this monthly electricity consumption by 12. This method is termed as "Summer Months Method (SMM)".

---

[13] It was assumed that the DHW was distributed at 45°C (318K) in the households. Therefore, ΔT used in Equation C.1 is 318K- 285K = 33K.

It was assumed that the monthly DHW consumption is constant throughout the year. Thus, the month-to-month variation in the DHW heating energy consumption would be due to the month-to-month change in the average ground temperature. It was also assumed that the DHW temperature is maintained at 318 K (45°C). For example, the DHW heating energy consumption (Q) of a household in January and in August would be:

$$Q_{JAN} = m[kg] * c[MJ/kgK] * (318 - GT_{JAN}[K])$$

$$Q_{AUG} = m[kg] * c[MJ/kgK] * (318 - GT_{AUG}[K])$$

where GT is the average annual ground temperature in K. The ratio of the DHW heating energy consumption in January to that in August is:

$$\frac{Q_{JAN}}{Q_{AUG}} = \frac{318 - GT_{JAN}}{318 - GT_{AUG}} = \frac{\Delta T_{JAN}}{\Delta T_{AUG}}$$

If the household's electricity consumption in August was the minimum monthly electricity consumption within summer months, *i.e.* there was no electricity consumption for SH in August, and electricity consumption in this month accounted for ALC and DHW heating. Then, the DHW heating energy consumption in January can be calculated applying the SMM as:

$$Q_{JAN} = Q_{AUG} * \frac{318 - GT_{JAN}}{318 - GT_{AUG}} = Q_{AUG} \frac{\Delta T_{JAN}}{\Delta T_{AUG}}$$

Therefore, the SMM can be formulized as:

$$Q_i = \frac{Q_j}{\Delta T_j} \Delta T_i \qquad\qquad (C.3)$$

where Q is the monthly electricity consumption, ΔT is the temperature difference (*i.e.* 318 – GT), i represents the months other than the month with the minimum electricity consumption, and j represents the month with the minimum electricity consumption.

The monthly electricity consumption information of the 531 households was required to apply the SMM. Out of 531 households, 525 of them had monthly electricity bills as given in Table B.3 of Appendix B. The average monthly ground temperatures were available only for 22 cities in 1993 from Environment Canada (Environment Canada, 1999). The cities, in which these 525 households with monthly electricity bills were located, were matched with the closest cities with the average monthly ground temperature data.

175

The electricity consumption data of the 525 households using electricity for DHW and SH were screened before applying the SMM.

1. There were 75 households with A/C units in the set of 525 households with monthly electricity bills. The A/C unit increases the electricity consumption in the summer months, therefore these 75 households were excluded from the analysis.

2. The monthly electricity consumption values of each of the remaining 450 (=525-75) households were plotted to check if the minimum electricity consumption occurred in one of the three summer months and if the household had a "reasonable" annual electricity consumption pattern. An electricity consumption pattern was deemed reasonable if the peak consumption occurs in winter months and the minimum occurs in summer months. The monthly electricity consumption of a sample household using electricity for ALC, and DHW and SH with a "reasonable" annual electricity consumption pattern, and the monthly average ground temperature at the region the household was located are given in Figure C.1. As seen in Figure C.1, the household had a "reasonable" annual electricity consumption pattern, which gradually increases in winter months and decreases in summer months. It was found that 294 households did not have "reasonable" electricity consumption patterns; such as minimum electricity consumption occurring in one of the winter months, or maximum electricity consumption occurring in one of the summer months, or sudden increase or decrease in consumption occurring in-between seasons. Thus, these 294 households were excluded from the analysis.

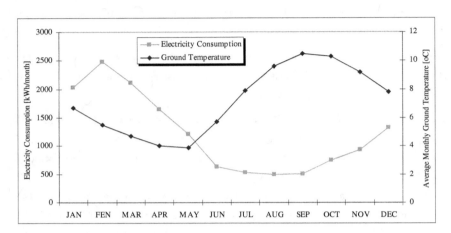

Figure C.1. Monthly electricity consumption and average monthly ground temperature of a sample household using electricity for DHW and SH

The annual ALC electricity consumption estimated by the ALC NN model for these households was divided by twelve and deducted from each month's electricity consumption of the remaining 156 (= 450-294) households. Then, the SMM was applied to the each of 156 households as follows:

- Electricity consumption of each month was divided by the monthly temperature difference of that month, *i.e.* ΔT in Equation C.3 (= 318 – monthly GT).

- The monthly electricity consumption with the minimum $\dfrac{Q}{\Delta T}$ value was taken as the month in which there was no electricity consumption for SH, and the household electricity consumption accounted only for DHW heating. This value was then multiplied by the monthly temperature difference of each of the remaining eleven months.

- The summation of computed Q values of the eleven months and the electricity consumption of the month with minimum $\dfrac{Q}{\Delta T}$ value represented the annual DHW heating electricity consumption.

The annual DHW heating electricity consumption values of the 156 households estimated by the SMM were analyzed, and it was found that:

1. There were 40 households with negative DHW heating electricity consumption values. Thus, these 40 households were excluded from the analysis.

2. There were also 26 households with DHW heating consumption values less than 1,745 kWh. These 26 households were excluded from the analysis.

Thus, the number of households with electricity bills and using electricity for ALC, and space and DHW heating reduced to 90 after excluding the 66 (= 40 + 26) households as given above.

As a total, 431 (= 341 + 90) households with electricity bills remained in the dataset. An additional screening process was applied to the remaining 436 households by examining their daily DHW consumption values.

As the number of occupants increases, the daily DHW consumption increases. The household with the maximum number of occupants in the 1993 SHEU database have eleven occupants. Therefore, this household would likely have the maximum daily DHW consumption amongst the households in the 1993 SHEU database. Using Equation C.2, it was found that the daily DHW consumption of this household was 470 l/day.

The maximum number of weekly dishwasher and clothes washer loads reported in the 1993 SHEU database was 15 loads per week. It was stated by Tomlimson and Rizy (1998) that average clothes washer DHW consumption was 25 liters per load in a study covering 103 households in USA. The average dishwasher DHW consumption was reported in EnerGuide- 1994 (NRCan, 1994) as 40 liters for one load. Therefore, daily dishwasher and clothes washer DHW consumption of a household with 15 loads per week would be about 140 liters.

Thus, the maximum likely daily DHW consumption for the households in the 1993 SHEU database would be 610 l/day. Therefore, the daily DHW consumption values of the remaining 431 households with electricity bills should be lower than this value.

The DHW consumption values of the remaining 431 households were calculated using Equation C.4.

$$m = \frac{Q * EF}{365\,[\text{d/yr}] * c * \Delta T * \rho} \tag{C.4}$$

where m : daily DHW consumption, L/day

Q : DHW heating energy consumption, MJ/yr

EF : efficiency (0.96 for tankless systems and 0.824 for systems with tanks)

178

c : 4.184 10^{-3} MJ/kgK

ΔT : 318 K – Average annual ground temperature, K

ρ : water density, kg/L = 0.997 kg/L

It was found that there were 43 households with DHW consumption values higher than 610 L/day. These 43 households were excluded from the analysis, and the remaining 388 (= 431 – 43) households with electricity bills were deemed to have reliable electricity consumption data to be used in the dataset for the development of the DHW network.

2. Households with Natural Gas Bills

There are 1,012 households with natural gas bills. The distribution of these households based on their end-use natural gas consumption is given in Table B.4 of Appendix B. As seen in Table B.4 of Appendix B, there are 27 households using natural gas only for DHW heating, and 915 households using natural gas for DHW and SH. There are seven households with natural gas bills, but the data indicate that they do not use natural gas fueled space or DHW heating equipment, or any other natural gas fueled appliances.

Out of 915 households using natural gas for space and DHW heating, 745 of them had monthly natural gas bills as given in Table B.5 of Appendix B. Therefore, the annual DHW heating natural gas consumption of these 745 households with monthly natural gas bills could be estimated using the SMM. Before applying the SMM, some screening processes were applied to these 745 households:

1. The monthly natural gas consumption of each of the 745 households was examined if the minimum monthly natural gas consumption occurred in the summer months (*i.e.* June, July, and August). It was found that 488 households had minimum monthly natural gas consumption in January, February, March, April, May, September, October, November, or December. Thus, these 488 households were excluded from the analysis.

2. The minimum likely annual DHW heating natural gas consumption was calculated using Equation C.1 and C.2, and assuming that:
   - the household has one occupant,
   - DHW heating system efficiency is 0.80,
   - average annual ground temperature is 12 °C (285K).

179

With these assumptions, the minimum likely annual DHW heating natural gas consumption was calculated as 7,536.8 MJ/yr $\approx$ 202 m^3/yr $\approx$ 17 m^3/month. There were three households with monthly DHW heating consumption values less than 17 m^3. These three households were excluded from the analysis.

3. There might be other sources of natural gas consumption in the households other than the natural gas fueled space and DHW heating systems. The natural gas fueled appliances reported in the 1993 SHEU database that would contribute to the natural gas consumption of a household are:
   - Oven/stove
   - Dryer
   - Pool heater
   - Fire place
   - Supplementary heater

   There were 59 households with monthly natural gas bills using one or more of the above natural gas fueled appliances. Thus, these 59 households were excluded from the analysis.

4. The monthly natural gas consumption values of each of the remaining 195 (= 745 – 488 – 3 – 59) households were plotted to check if the household had a "reasonable" annual natural gas consumption pattern, which gradually increases in winter months and decreases in summer months. It was found that 26 households did not have "reasonable" monthly natural gas consumption patterns. Thus, these 26 households were excluded from the analysis.

Thus, the number of households with monthly natural gas bills reduced to 169 after excluding the 576 (488+3+59+26) households as given above. The SMM was applied to these remaining 169 households, and their annual DHW heating natural gas consumption was estimated.

There were 27 households using natural gas only for DHW heating, 19 of these households had natural gas consumption values either less than 200 m^3 or more than 2000 m^3. Thus, 19 households with unacceptable DHW heating energy consumption values were excluded from the analysis. Therefore, only eight of these 27 households were deemed to have acceptable natural gas DHW heating consumption values.

As a total, 177 (= 169 + 8) households with natural gas bills were left in the analysis. An additional screening process was applied to the remaining 177 households by examining their daily DHW consumption values. The daily DHW consumption values of these 177 households were calculated using Equation C.4. Two households with daily DHW consumption values higher than

180

610 L/day were excluded from the analysis. Finally, remaining 175 (= 177 – 2) households with natural bills were deemed to have reliable natural gas consumption data used to be used in the dataset for the development of the DHW network.

3. Households with Oil Bills

There are 236 households with oil bills. The distribution of these households based on their end-use consumption is given in Table B.6 of Appendix B. As seen in Table B.6 of Appendix B, there are three households using oil only for DHW heating and 137 households using oil for DHW and SH. There are six households with oil bills, but the data indicate that they do not use oil fueled space or DHW heating equipment, or any other oil fueled appliances.

The billing data of the 230 households with oil bills were all in the annual form. Since there was no household with monthly oil bills, the SMM could not be applied to the households with oil bills and using oil for space and DHW heating.

There were three households with oil bills and using oil only for DHW heating. The annual oil consumption values of these three households were all higher than 2700 L. The maximum likely annual oil consumption for DHW heating was calculated using Equation C.1 and C.2, and assuming that:
- the household has eleven occupants (maximum number of occupants reported in the 1993 SHEU database),
- DHW heating system efficiency is 0.50,
- average annual ground temperature is 5 °C (278K, minimum average annual ground temperature in Canada in 1993).

With these assumptions, the maximum likely annual DHW heating oil consumption was calculated as 57,250 MJ/yr ≈ 1500 L/yr. Therefore, the DHW heating oil consumption values of these three households were found unacceptably high. Thus, none of the 230 households oil bills were found reliable to be used in the dataset for the development of the DHW network.

APPENDIX D

DEVELOPMENT OF THE SH NETWORK DATASET

SH NETWORK DATASET

Billing data obtained from fuel suppliers and utility companies were available for a subset of 3,341 households in the 1993 SHEU database. As shown in Appendix B, a total of 3,298 bills from 2,749 households were considered reliable. Out of these 2,749 households, there are 2,050 households with electricity bills, 1,012 households with natural gas bills, and 236 households with oil bills. The screening processes used in the selection of the households with the billing data for the development of the SH network dataset are presented in the following sections.

1. Households with Electricity Bills

There are 2,050 households with electricity bills. The distribution of these households based on their end-use electricity consumption is given in Table B.2 of Appendix B. As seen in Table B.2 of Appendix B, there are 25 households using electricity for ALC and SH, and 531 households using electricity for ALC, DHW, and SH.

The ALC electricity consumption estimated by the ALC NN model was deducted from the annual electricity consumption of the 25 households which use electricity for ALC and SH, and 12 households were found to have negative values. Thus, these 12 households were excluded from the analysis.

There are 531 households using electricity for ALC, DHW, and SH in the billing dataset. The ALC and DHW NN models were used to estimate the ALC and DHW heating electricity consumption of these households, respectively. The SH electricity consumption of the 531 households were calculated by deducting the ALC and DHW heating electricity consumption estimates from the annual electricity consumption. From this calculation, 55 households were found to have negative values. Thus, these 55 households were excluded from the analysis.

The number of households with electricity bills reduced to 489 after excluding 67 (= 55 + 12) households as explained above. The distribution of the electricity billing data of the 489 households is given in Figure D.1.

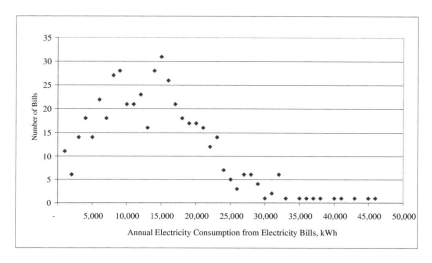

Figure D.1. Distribution of the electricity billing data of the 489 households

As seen from Figure D.1, the annual electricity consumption values of the 489 households are mostly between 5,000 kWh and 35,000 kWh. Thus, the 72 households with billing data less than 5,000 kWh or more than 35,000 kWh were excluded from the analysis. This left 417 households with electricity billing data in the analysis.

2. Households with Natural Gas Bills

There are 1,012 households with natural gas bills. The distribution of these households based on their end-use natural gas consumption is given in Table B.4 of Appendix B. As seen in Table B.4 of Appendix B, there are 63 households using natural gas only for SH, and there are 915 households using natural gas for DHW and SH. There are seven households with natural gas bills, but the data indicate that they do not use natural gas fueled space or DHW heating equipment, or any other natural gas fueled appliances.

There might be other sources of natural gas consumption in the households with natural gas bills other than the natural gas fueled space and DHW heating systems. The natural gas fueled appliances reported in the 1993 SHEU database that would contribute to the natural gas consumption of a household are:

- Oven/stove

- Dryer
- Pool heater
- Fire place
- Supplementary heater

There were 165 households with natural gas bills using one or more of the above natural gas fueled appliances. Thus, these 165 households were excluded from the analysis. This reduced the number of households with natural gas bills to 813: 53 of them use natural gas only for SH and 760 of them use natural gas for DHW and SH.

The DHW heating natural gas consumption of the 760 households was estimated by the DHW NN model. The SH natural gas consumption of these 760 households were calculated by deducting the DHW heating natural gas consumption estimates from the annual natural gas consumption. The distribution of the natural gas billing data of the 813 households is given in Figure D.2.

As seen from Figure D.2, the annual natural gas consumption values of the 813 households are mostly between 700 m^3 and 5,000 m^3. Thus, the 29 households with billing data less than 700 m^3 or more than 5,000 m^3 were excluded from the analysis. This left 784 households with natural gas billing data in the analysis.

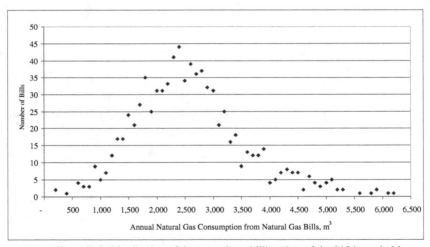

Figure D.2. Distribution of the natural gas billing data of the 813 households

185

3. Households with Oil Bills

There are 236 households with oil bills. The distribution of these households based on their end-use consumption is given in Table B.6 of Appendix B. As seen in Table B.6 of Appendix B, there are 90 households using oil only for SH and 137 households using oil for DHW and SH. There are six households with oil bills, but the data indicate that they do not use oil fueled space or DHW heating equipment, or any other oil fueled appliances.

The DHW heating energy consumption of the households using oil were not estimated by the DHW NN model, since the households using oil for DHW heating were not included in the dataset used to develop the DHW NN model. Therefore, the 137 households using oil for space and DHW could not be included in the dataset since their oil consumption for DHW heating were not estimated.

There might be other sources of oil consumption in the 90 households with oil bills other than the oil fueled SH systems. The oil fueled appliances reported in the 1993 SHEU database that would contribute to the oil consumption of a household are:

- Oven/stove
- Pool heater
- Fire place
- Supplementary heater

There were three households with oil bills using one or more of the above oil fueled appliances. Thus, these three households were excluded from the analysis. This reduced the number of households with oil bills to 87. The distribution of the oil billing data of the 87 households is given in Figure D.3.

186

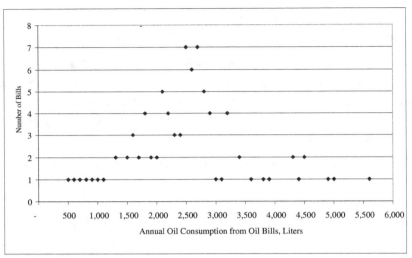

Figure D.3. Distribution of the oil billing data of the 87 households

As seen from Figure D.3, the annual oil consumption values of the 87 households are mostly between 700 L and 4,500 L. Thus, the six households with billing data less than 700 L or more than 4,500 L were excluded from the analysis. This left 81 households with oil billing data in the analysis.

4. Final SH NN Model Dataset

These exclusions left 1,282 households in the analysis: 417 with electricity bills, 784 with natural gas bills, and 81 with oil bills. During the input unit selection step, it was found that 43 households in the dataset have missing information for the chosen input units in the 1993 SHEU database. Thus, these 43 households were excluded from the analysis.

A few number of households in the database use heat pump systems as their SH equipment. Due to the small number of households with these systems and lack of information on the coefficient of performance values of these systems, eleven households that use heat pump systems in the dataset were excluded from the analysis. Thus, the number of households left in the dataset is 1,228 (1282 - 43 –11): 396 with electricity bills, 755 with natural gas bills, and 77 with oil bills.

APPENDIX E

CONFIGURATION OF THE NN MODEL

1. ALC NN MODEL

The NN model developed to estimate the ALC energy consumption has 55 input layer units, nine units in each of the three hidden layers, and one output layer unit. As shown in Figure E.1, the units from one to 55 represent the input layer units, units from 56 to 82 represent the hidden layer units, and units 83 is the output layer unit.

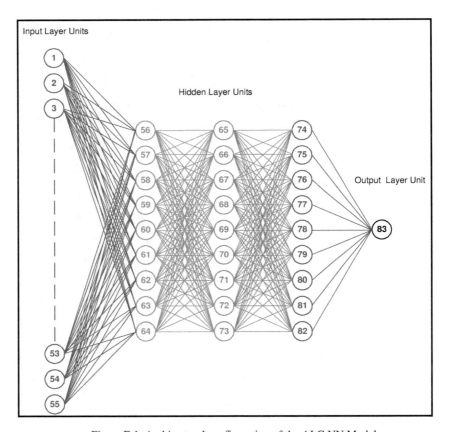

Figure E.1. Architectural configuration of the ALC NN Model

The unit definition section of the SNNS output for the ALC NN Model is given in Figure E.2. The bias values of the hidden and output layer units are given in the third column.

189

No.	Name	Bias	Type	Activation F.
1	Main Refrigerator		input	
2	Secondary Refri.		input	
3	Stove		input	
4	Dishwasher		input	
5	Main Freezer		input	
6	Secondary Freezer		input	
7	Clothes Washer		input	
8	Clothes Dryer		input	
9	Microwave		input	
10	Color TV		input	
11	BW TV		input	
12	Furnace Fan		input	
13	Boiler Pump		input	
14	Supplementary Ht.		input	
15	VCR		input	
16	CD Player		input	
17	Stereo		input	
18	Computer		input	
19	Electric Blanket		input	
20	Water Bed		input	
21	Humidifier		input	
22	Dehumidifier		input	
23	Car Block		input	
24	Car Warmer		input	
25	Water Cooler		input	
26	Fish Tank		input	
27	Bath. Exh. Fan		input	
28	Kitch. Exh. Fan		input	
29	Central Air Filter		input	
30	Central Humidifi.		input	
31	Central Dehumidif.		input	
32	Central Air Exch.		input	
33	HRV		input	
34	Central Vacuum		input	
35	Sump Pump		input	
36	Water Softener		input	
37	Jacuzzi		input	
38	Sauna		input	
39	Ceiling Fan		input	
40	Portable Fan		input	
41	Central A/C		input	
42	Window A/C		input	
43	Halogen		input	
44	Fluorescent		input	
45	Incandescent		input	
46	HDD		input	
47	CDD		input	
48	Area		input	
49	Income		input	
50	Dwelling Type		input	

Figure E.2. Unit definition of the ALC NN Model

```
51 | Ownership          |          | input  |          |
52 | Population         |          | input  |          |
53 | No. of Children    |          | input  |          |
54 | No. of Adults      |          | input  |          |
55 | Employment Ratio   |          | input  |          |
56 |                    | -0.03352 | hidden | Logistic |
57 |                    |  0.43477 | hidden | Logistic |
58 |                    |  0.54654 | hidden | Logistic |
59 |                    | -0.14504 | hidden | Logistic |
60 |                    | -0.58060 | hidden | Logistic |
61 |                    | -0.99884 | hidden | Logistic |
62 |                    |  0.98134 | hidden | Logistic |
63 |                    | -0.89134 | hidden | Logistic |
64 |                    | -1.36429 | hidden | Logistic |
65 |                    | -0.77716 | hidden | Logistic |
66 |                    |  0.10899 | hidden | Logistic |
67 |                    | -0.62399 | hidden | Logistic |
68 |                    |  0.85723 | hidden | Logistic |
69 |                    |  0.93863 | hidden | Logistic |
70 |                    |  0.09220 | hidden | Logistic |
71 |                    | -0.59255 | hidden | Logistic |
72 |                    | -0.74478 | hidden | Logistic |
73 |                    | -1.17445 | hidden | Logistic |
74 |                    | -0.77444 | hidden | Logistic |
75 |                    | -0.07289 | hidden | Logistic |
76 |                    |  0.28001 | hidden | Logistic |
77 |                    |  0.03338 | hidden | Logistic |
78 |                    | -0.24248 | hidden | Logistic |
79 |                    |  0.00637 | hidden | Logistic |
80 |                    | -0.53576 | hidden | Logistic |
81 |                    | -0.72210 | hidden | Logistic |
82 |                    | -0.59786 | hidden | Logistic |
83 |                    | -0.29669 | output | Identity |
----|--------------------|----------|--------|--------------|
```

Figure E.2. (continued) Unit definition of the ALC NN Model

The values of the weights between the units of the ALC NN Model are given in Figure E.3.

```
target | site | source
-------|------|-------

56 |  1: 0.48256,  2: 0.99136,  3:-0.15941,  4: 0.62603,  5: 0.18473,  6: 0.88864,  7: 2.31364,  8: 0.66008,  9:-0.64953,
   | 10: 0.32304, 11:-0.43343, 12: 1.24811, 13:-0.77258, 14: 2.27300, 15: 0.40797, 16:-0.94207, 17:-0.53683, 18: 0.26898,
   | 19: 0.54584, 20: 0.32996, 21: 0.43003, 22:-0.30338, 23:-0.42901, 24:-0.12919, 25:-0.64250, 26: 0.22126, 27:-0.62180,
   | 28:-2.01902, 29: 1.19463, 30:-0.98469, 31: 0.15268, 32:-0.03429, 33: 0.85540, 34:-0.90635, 35:-0.61286, 36:-1.22549,
   | 37: 0.25483, 38:-0.37178, 39:-0.51672, 40: 0.18780, 41: 0.78861, 42: 0.03183, 43: 0.55878, 44:-0.68991, 45: 2.22690,
   | 46:-0.60508, 47:-1.35957, 48: 0.97104, 49: 1.73880, 50: 0.53295, 51: 0.39831, 52:-4.69737, 53:-0.85387, 54:-0.54590,
   | 55:-3.30050

57 |  1:-0.19028,  2: 0.44840,  3: 0.30907,  4:-2.39816,  5:42.78911,  6: 1.05165,  7:27.62451,  8:20.10308,  9: 0.09716,
   | 10:-0.39381, 11: 0.70016, 12:-6.72592, 13: 0.92062, 14: 0.21791, 15: 0.06547, 16: 1.65947, 17: 0.59692, 18:-0.78709,
   | 19: 0.35367, 20: 0.63229, 21: 0.62519, 22:-0.22147, 23: 0.93994, 24:-0.51080, 25: 0.99098, 26: 0.17166, 27:-1.42409,
   | 28:-21.75667, 29:-0.75556, 30: 9.58150, 31:-0.68399, 32:-13.48908, 33:-0.79857, 34: 0.53539, 35: 9.08046, 36: 3.27121,
   | 37:-9.56089, 38:-0.59009, 39:-0.43419, 40:-0.23955, 41:-0.42737, 42:-0.09456, 43: 0.50683, 44:-0.82556, 45:-2.39379,
   | 46: 5.20302, 47: 0.29707, 48:-0.19991, 49:-2.47678, 50: 0.21441, 51:-0.35103, 52: 1.94220, 53:-0.79658, 54: 0.76340,
   | 55: 0.45616

58 |  1:-0.53811,  2: 0.30621,  3: 0.38221,  4: 0.57114,  5: 0.30353,  6: 0.47719,  7: 1.14937,  8: 0.64917,  9:-0.32991,
   | 10: 0.80031, 11: 0.40676, 12:-0.15867, 13: 0.31201, 14:-0.59390, 15: 0.66684, 16:-0.40148, 17:-0.68761, 18:-0.76055,
   | 19:-0.32377, 20: 0.84475, 21:-0.80545, 22: 0.47911, 23:-0.23517, 24:-0.70883, 25:-0.68291, 26: 0.23114, 27:-0.53076,
   | 28: 0.15832, 29:-0.24364, 30: 0.12129, 31:-0.74294, 32:-0.22998, 33:-0.50808, 34: 0.77865, 35:-0.80386, 36:-1.41192,
   | 37: 0.74343, 38:-0.97911, 39: 0.87674, 40:-0.27208, 41: 0.93049, 42: 0.22388, 43:-0.07016, 44: 0.44619, 45: 0.03339,
   | 46:-0.38579, 47:-1.0221, 48: 0.09807, 49:-5.83082, 50: 0.07508, 51: 0.45039, 52: 6.49628, 53: 0.59759, 54: 1.00058.
   | 55: 1.52997

59 |  1: 0.91906,  2:-0.18460,  3: 0.76515,  4: 0.55644,  5: 0.13095,  6:-0.46788,  7: 0.18855,  8:-0.48709,  9: 0.61552,
   | 10: 0.76980, 11: 0.30157, 12: 0.25421, 13:-0.31260, 14: 0.78390, 15: 3.62706, 16:-0.71080, 17: 0.93177, 18: 0.44440,
   | 19: 0.21319, 20:-0.05112, 21: 0.38023, 22:-0.63762, 23: 0.12323, 24: 0.82067, 25: 0.48184, 26:-0.83276, 27:-0.86953,
   | 28: 0.35835, 29: 1.52933, 30: 0.98086, 31: 0.25448, 32:-0.50296, 33: 0.08019, 34:-0.76064, 35: 0.21838, 36:-0.21500,
   | 37:-0.86115, 38: 0.30130, 39: 0.14469, 40: 0.63669, 41: 1.97370, 42: 0.32503, 43: 0.09314, 44: 1.07737, 45:-0.53937,
   | 46: 1.10545, 47:-5.71166, 48: 0.39250, 49: 1.30304, 50: 0.15173, 51: 0.93614, 52:-7.25469, 53:-0.34511, 54:-0.29782,
   | 55: 2.69671

60 |  1:-0.51888,  2:-2.60096,  3:-0.32683,  4:-1.00154,  5:-0.63978,  6: 0.01819,  7: 2.99153,  8:-0.63305,  9: 0.78810,
   | 10:-0.44078, 11: 0.26846, 12:-0.99722, 13: 0.44442, 14: 0.05902, 15:-1.21324, 16:-0.37208, 17:-0.84703, 18: 0.36363,
   | 19: 0.27363, 20:-0.92052, 21:-0.20120, 22:-0.60489, 23: 0.18210, 24:-0.51052, 25: 0.94696, 26: 0.80900, 27:-0.05915,
   | 28:-1.52926, 29:-0.93684, 30: 1.11394, 31: 0.24873, 32:-0.64783, 33: 0.52551, 34: 0.97483, 35: 0.66110, 36:-0.39279,
   | 37: 0.38877, 38:-0.53956, 39:-0.44312, 40: 0.25231, 41:-0.03606, 42: 0.39181, 43: 0.23684, 44: 0.94220, 45:-3.76983,
   | 46: 3.27561, 47: 0.65145, 48:-0.35282, 49:-5.41727, 50:-0.58438, 51: 0.67349, 52: 3.00446, 53:-0.82996, 54:-0.59494.
   | 55: 2.36854

61 |  1: 0.11610,  2: 1.96994,  3:-0.21475,  4: 0.98495,  5:446.91061,  6: 0.74287,  7:276.18372,  8: 2.22961,  9:-0.24725,
   | 10:113.76408, 11:-0.57393, 12:-292.17062, 13: 0.16124, 14:-0.88170, 15:-0.67514, 16: 0.07705, 17:-0.66849, 18: 0.46629,
   | 19: 0.34535, 20:-0.48462, 21: 0.94068, 22: 0.50499, 23: 0.19130, 24:-0.02282, 25:-0.75741, 26:-0.10921, 27:-0.40585,
   | 28:-432.40007, 29: 0.63351, 30:585.86255, 31: 0.94675, 32: 0.53953, 33: 0.67308, 34: 1.48365, 35: 0.67716, 36: 5.04117,
   | 37: 0.86751, 38: 0.72921, 39:-0.71366, 40:-0.55527, 41: 1.81204, 42: 0.17775, 43:-0.93343, 44:-0.41110, 45: 0.20248,
   | 46: 0.80785, 47: 0.56661, 48:-0.46647, 49:-4.89840, 50:-0.53808, 51: 0.54491, 52: 0.27194, 53: 0.15940, 54:60.07847,
   | 55:80.97397
```

Figure E.3. Values of the weights between the units of the ALC NN Model

62
1:-0.07242, 2:-0.60115, 3: 0.87054, 4: 1.02072, 5: 1.17484, 6: 0.38725, 7: 1.78980, 8: 1.55850, 9:-0.65180,
10: 0.41986, 11: 0.80738, 12:-0.27021, 13:-0.27021, 14: 0.92397, 15:-0.19391, 16: 0.62715, 17: 0.76027, 18: 0.58186,
19:-0.31932, 20: 0.98173, 21: 0.43502, 22:-0.61960, 23: 0.99039, 24:-0.85083, 25: 0.24868, 26:-0.16814, 27:-0.32083,
28: 2.04940, 29: 0.72066, 30:-0.07931, 31: 0.63940, 32: 0.69447, 33: 0.57092, 34: 0.42728, 35:-0.69541, 36: 0.09148,
37: 0.09444, 38:-0.67710, 39:-0.61323, 40: 0.89241, 41: 1.14071, 42: 0.13251, 43:-0.49913, 44: 1.00538, 45: 0.44192,
46:-4.49438, 47:-6.233362, 48:-0.276646, 49: 1.30358, 50:-0.13662, 51: 0.84855, 52:-1.009191, 53:-0.01322, 54: 0.42936,
55: 0.98292

63
1: 0.24600, 2: 0.91154, 3:-0.40082, 4: 0.64718, 5: 1.44811, 6:-0.74746, 7: 8.12743, 8: 1.94892, 9: 5.19863,
10: 1.88677, 11: 0.27565, 12:-0.15655, 13:-0.92379, 14: 6.55690, 15: 8.83653, 16:-0.57692, 17: 0.90165, 18:-0.86714,
19:-0.36538, 20: 0.45777, 21:-0.43145, 22:-0.80454, 23:-0.28233, 24:-0.82421, 25:-0.05488, 26:-0.95065, 27:-0.54893,
28: 1.63376, 29: 0.19474, 30:-0.55210, 31: 0.60473, 32:-0.86694, 33: 0.70207, 34:-2.73702, 35:-2.25123, 36:-1.90679,
37: 0.60300, 38: 0.98717, 39: 0.25927, 40: 0.24240, 41: 2.24685, 42:-0.08721, 43:-0.34663, 44:-0.12275, 45: 5.70941,
46:-11.50117, 47:-1.06291, 48:-0.025567, 49: 3.05491, 50: 0.16423, 51:-0.87730, 52:-0.04814, 53: 0.01358, 54: 0.06384,
55: 0.18087

64
1:-0.08912, 2: 0.99841, 3:-0.25886, 4: 0.72362, 5: 0.46511, 6: 0.51784, 7: 0.67411, 8: 1.23101, 9: 0.53343,
10: 0.19912, 11:-0.08284, 12: 1.63068, 13: 0.12645, 14:-0.13703, 15: 0.42866, 16: 0.88001, 17:-0.69169, 18:-0.16926,
19: 0.86554, 20:-0.29513, 21:-0.20867, 22: 0.83432, 23:-0.37343, 24: 0.75765, 25: 0.69702, 26:-0.12705, 27:-0.82925,
28:-0.74718, 29:-0.43568, 30:-0.46866, 31:-0.77629, 32:-0.22392, 33:-0.15858, 34:-0.16788, 35:-0.66829, 36:-0.19025,
37:-0.84529, 38: 0.29580, 39: 0.35259, 40:-0.14276, 41: 0.11916, 42:-0.45407, 43: 0.05459, 44:-0.21635, 45: 0.05213,
46:-1.12080, 47:-3.74273, 48: 0.68902, 49: 0.10143, 50: 0.05884, 51: 0.89419, 52: 1.33886, 53: 0.48624, 54: 4.21965,
55: 0.73063

65 56: 0.63607, 57: 1.64477, 58:-0.09972, 59: 0.26587, 60:-0.78649, 61: 0.00674, 62: 1.03045, 63: 1.01646, 64: 0.56886,
66 56: 0.28392, 57: 1.65500, 58:-1.76230, 59: 0.70067, 60:-0.43721, 61:-9.95513, 62: 2.00668, 63:-1.92340, 64: 0.52268,
67 56: 0.55094, 57:-0.43308, 58: 0.01858, 59: 1.20955, 60:-0.61394, 61:-1.72432, 62: 0.72638, 63:-0.15532, 64:-0.03068,
68 56:-0.54127, 57:-0.23087, 58:-0.93297, 59:-0.18466, 60: 0.66420, 61: 0.27362, 62:-0.25447, 63:-0.96173, 64:-1.46426,
69 56: 0.18172, 57: 0.98936, 58: 0.54577, 59: 0.79883, 60: 0.47602, 61:-0.59835, 62:-0.46974, 63:-0.23372, 64:-0.87014,
70 56:-0.74877, 57: 0.02402, 58:-0.78146, 59: 0.76451, 60: 0.45890, 61:-0.08539, 62:-1.00785, 63:-1.04649, 64: 0.79857,
71 56: 0.33562, 57:-0.95560, 58:-0.45851, 59: 0.15283, 60: 0.73723, 61:-3.62504, 62: 0.22837, 63:-0.15983, 64:-10.31943,
72 56: 0.44428, 57:-2.51379, 58:-0.88572, 59:-1.08168, 60:-0.96700, 61:-14.36240, 62:-1.17265, 63: 1.61733, 64:-0.19030,
73 56:-0.46577, 57: 1.13745, 58: 0.44062, 59:-1.02865, 60: 0.61286, 61:-0.52541, 62: 0.42849, 63:-0.22314, 64:-0.60129,
74 65: 0.25067, 66: 0.60164, 67:-0.21111, 68:-0.23053, 69: 0.83320, 70:-0.10702, 71: 0.24360, 72: 0.68483, 73: 0.71612,
75 65: 0.79068, 66: 0.42122, 67:-0.33296, 68: 0.83914, 69: 0.38410, 70: 0.27356, 71:-0.62772, 72:-1.05171, 73: 0.59846,
76 65: 0.03925, 66: 0.29766, 67: 0.81809, 68: 0.24879, 69: 0.65074, 70:-0.13281, 71:-0.87193, 72: 0.39426, 73:-0.00161,
77 65: 0.14313, 66:-0.73094, 67:-0.20339, 68:-1.07876, 69:-1.04320, 70:-0.06285, 71:-0.35624, 72:-0.44609, 73:-0.22495,
78 65:-0.51008, 66: 0.53066, 67:-0.82659, 68: 0.61907, 69: 0.27559, 70:-0.05542, 71: 0.53932, 72: 0.53481, 73:-0.96357,
79 65:-0.34649, 66:-0.38600, 67: 0.35350, 68: 0.00477, 69: 0.80559, 70: 0.56182, 71: 0.34008, 72: 0.47372, 73:-0.73538,
80 65: 0.57641, 66:-0.44815, 67:-0.07576, 68:-0.58934, 69:-0.39963, 70: 0.00276, 71:-0.68686, 72: 0.73423, 73:-0.62607,
81 65:-0.64508, 66:-0.59118, 67:-0.06696, 68:-0.60060, 69:-0.74603, 70:-0.21230, 71:-0.98984, 72:-0.95701, 73: 0.46247,
82 65:-0.03206, 66: 1.32848, 67:-0.20031, 68:-0.82768, 69:-0.70676, 70:-0.27269, 71:-0.32752, 72: 0.70583, 73: 0.07952,
83 74: 0.58603, 75: 0.73478, 76:-0.26153, 77: 0.31178, 78:-0.82135, 79:-0.15511, 80: 0.29507, 81:-0.62470, 82: 0.20779

Figure E.3. (continued) Values of the weights between the units of the ALC NN Model

193

The output of the ALC NN Model can be calculated by applying Equations E.1 to E.4. Since the hidden layer activation function was chosen to be the logistic function (Equation 2.2), the outputs of the hidden layer units are calculated using the logistic function. Identity function (Equation 2.4) was chosen for the output layer unit, thus the output of the unit at the output layer is calculated using the identity function.

The output of each first hidden layer unit is calculated as follows:

$$z_j = \frac{1}{1 + e^{-\left(\sum\limits_{i=1}^{55} x_i v_{ij} + b_j\right)}} \qquad (E.1)$$

Similarly, the outputs of each second and third hidden layer units are calculated using Equations E.2 and E.3, respectively.

$$u_l = \frac{1}{1 + e^{-\left(\sum\limits_{j=56}^{64} z_j s_{jl} + b_l\right)}} \qquad (E.2)$$

$$r_k = \frac{1}{1 + e^{-\left(\sum\limits_{l=65}^{73} u_l t_{lk} + b_k\right)}} \qquad (E.3)$$

The output of the network is calculated as follows:

$$y_m = \sum\limits_{k=74}^{82} r_k w_{km} + b_m \qquad (E.4)$$

where,

x_i: input to the first hidden layer unit j from input layer unit i

z_j: output of the first hidden layer unit j / input to the second hidden layer unit l from first hidden layer unit j

u_l: output of the second hidden layer unit l / input to the third hidden layer unit k from second hidden layer unit l

r_k: output of the third hidden layer unit k / input to the output layer unit m from third hidden layer unit k

v_{ij}:   weight between the input layer unit i and first hidden layer unit j

s_{jl}:   weight between the first hidden layer unit j and second hidden layer unit l

t_{lk}:   weight between the second hidden layer unit l and third hidden layer unit k

w_{km}:   weight between the third hidden layer unit k and output layer unit m

b_j: bias of the first hidden layer unit j

b_l: bias of the second hidden layer unit l

b_k: bias of the third hidden layer unit k

b_m: bias of the output layer unit m

i: number of input layer units (1 to 55)

j: number of first hidden layer units (56 to 64)

l: number of second hidden layer units (65 to 73)

k: number of third hidden layer units (74 to 82)

m: output layer unit (83)

2. DHW NN MODEL

The NN model developed to estimate the DHW heating energy consumption has 18 input layer units, 29 hidden layer, and one output layer unit. As shown in Figure E.4, the units from one to 18 represent the input layer units, units from 19 to 47 represent the hidden layer units, and units 48 is the output layer unit.

195

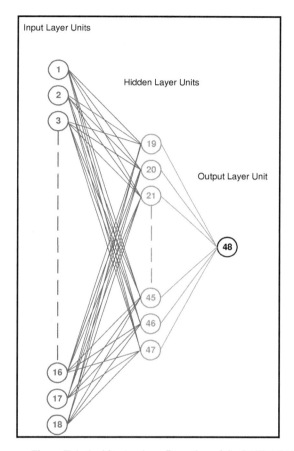

Figure E.4. Architectural configuration of the DHW NN Model

The unit definition section of the SNNS output for the DHW NN Model is given in Figure E.5. The bias values of the hidden and output layer units are given in the third column.

No.	Name	Bias	Type	Activation F.
1	Shared System		input	
2	No. of Tanks		input	
3	Tank Age		input	
4	Tank Size		input	
5	Tank Blanket		input	
6	Pipe Insulation		input	
7	Low Flow Showers		input	
8	Aerators		input	
9	System Efficiency		input	
10	Ground Temp.		input	
11	Clothes Washer Loads		input	
12	Dishwasher Loads		input	
13	No. of Children		input	
14	No. of Adults		input	
15	Income		input	
16	Dwelling type		input	
17	Ownership		input	
18	Population		input	
19		0.43783	hidden	Logistic
20		-0.19880	hidden	Logistic
21		0.47309	hidden	Logistic
22		0.26038	hidden	Logistic
23		-0.54827	hidden	Logistic
24		0.45152	hidden	Logistic
25		0.53175	hidden	Logistic
26		-0.27679	hidden	Logistic
27		0.34231	hidden	Logistic
28		0.46766	hidden	Logistic
29		0.87066	hidden	Logistic
30		0.56837	hidden	Logistic
31		0.06767	hidden	Logistic
32		0.27738	hidden	Logistic
33		0.06953	hidden	Logistic
34		-0.50163	hidden	Logistic
35		0.14777	hidden	Logistic
36		0.76758	hidden	Logistic
37		-0.75195	hidden	Logistic
38		-0.62799	hidden	Logistic
39		-0.81846	hidden	Logistic
40		-0.18931	hidden	Logistic
41		0.19161	hidden	Logistic
42		0.65300	hidden	Logistic
43		0.50905	hidden	Logistic
44		0.76856	hidden	Logistic
45		0.53077	hidden	Logistic
46		-0.83590	hidden	Logistic
47		-0.42768	hidden	Logistic
48		0.08354	output	Logistic

Figure E.5. Unit definition of the DHW NN Model

The values of the weights between the units of the DHW NN Model are given in Figure E.6

```
target | site | source:weight
-------|------|-------------------------------------------------------------------------------------------------------------------------------------------------------
  19   |      |  1: 0.06708,  2: 0.44278,  3:-0.51257,  4: 0.28401,  5: 0.61058,  6:-0.10674,  7:-0.65233,  8:-0.82151,  9:-0.56406,
       |      | 10:-0.64276, 11: 0.94141, 12: 0.11706, 13:-0.01500, 14:-0.51332, 15: 0.43779, 16:-0.91972, 17:-0.36780, 18:-0.24288
  20   |      |  1:-0.06959,  2:-0.74039,  3: 0.42109,  4: 0.72597,  5: 0.10609,  6:-0.02631,  7:-0.23414,  8:-0.22878,  9: 0.03116,
       |      | 10:-0.06126, 11: 0.09991, 12:-0.33456, 13: 0.29545, 14:-0.83966, 15: 0.29733, 16:-0.85876, 17:-0.93936, 18: 0.01421
  21   |      |  1:-0.19809,  2:-0.24634,  3: 0.62589,  4:-0.51699,  5: 0.66268,  6: 0.06115,  7: 0.39899,  8:-0.72256,  9: 0.24200,
       |      | 10:-0.40390, 11: 0.44581, 12:-0.75943, 13:-0.39686, 14: 0.05784, 15: 0.88829, 16:-0.53320, 17: 0.24903, 18: 0.58522
  22   |      |  1: 0.16097,  2:-0.28934,  3:-0.50660,  4:-0.28387,  5:-0.43035,  6: 0.48731,  7:-0.63953,  8:-0.70879,  9:-0.03176,
       |      | 10:-0.88879, 11: 0.12054, 12: 0.00886, 13: 0.84302, 14:-0.78123, 15:-0.79491, 16:-0.54350, 17: 0.17514, 18:-0.43127
  23   |      |  1:-0.20096,  2: 0.10222,  3: 0.05855,  4: 0.58591,  5:-0.70540,  6:-0.26732,  7: 0.93474,  8:-0.66794,  9: 1.15385,
       |      | 10:-0.65496, 11:-0.46845, 12: 0.30843, 13:-0.97725, 14: 0.85218, 15:-0.58545, 16:-0.03000, 17: 0.53455, 18:-0.13193
  24   |      |  1: 0.11947,  2: 0.51918,  3: 0.61313,  4: 0.94810,  5: 0.65985,  6: 0.31633,  7: 0.15938,  8:-0.37367,  9:-0.71083,
       |      | 10: 0.04237, 11: 0.83735, 12: 0.25570, 13: 0.09124, 14: 0.56347, 15: 0.53197, 16:-0.78222, 17: 0.29932, 18: 0.55718
  25   |      |  1: 0.03907,  2:-0.10781,  3:-0.28849,  4: 0.89287,  5:-0.04743,  6: 0.01979,  7:-0.58603,  8:-0.54097,  9: 0.24556,
       |      | 10:-0.50500, 11:-0.34287, 12: 0.80730, 13: 0.04404, 14: 0.15268, 15:-0.68155, 16:-0.35334, 17:-0.62920, 18:-0.36018
  26   |      |  1:-0.11791,  2:-0.21685,  3: 0.48417,  4:-0.63238,  5:-0.30580,  6: 0.00130,  7: 0.25857,  8:-0.18331,  9: 0.89399,
       |      | 10: 0.19961, 11:-0.25878, 12: 0.88619, 13: 0.66661, 14: 0.79577, 15:-0.43218, 16: 0.44224, 17:-0.77395, 18: 0.28588
  27   |      |  1: 0.33874,  2:-0.23464,  3: 0.44910,  4: 0.46670,  5: 0.48506,  6:-0.43770,  7: 0.68908,  8: 0.41161,  9:-0.33581,
       |      | 10: 0.34996, 11:-0.68438, 12: 0.43618, 13:-0.30993, 14:-0.61610, 15:-0.78602, 16:-0.25355, 17: 0.83514, 18:-0.83827
  28   |      |  1:-0.17437,  2:-0.72560,  3:-0.61096,  4: 0.30309,  5:-0.50100,  6: 0.36440,  7:-0.67213,  8:-0.07039,  9:-0.47677,
       |      | 10: 0.26802, 11:-0.36491, 12: 0.24718, 13:-0.83295, 14:-0.90795, 15: 0.78994, 16:-0.98827, 17: 0.86477, 18:-0.18838
  29   |      |  1: 0.02283,  2: 0.13784,  3: 0.38307,  4:-0.16934,  5: 0.04782,  6:-0.12043,  7: 0.05649,  8:-0.32136,  9:-0.18210,
       |      | 10: 0.37712, 11: 0.36147, 12: 0.55556, 13: 0.70691, 14: 0.18284, 15:-0.17974, 16: 0.73736, 17:-0.47060, 18: 0.60984
  30   |      |  1:-0.28017,  2:-0.52829,  3:-0.56923,  4:-0.41387,  5: 0.27946,  6: 0.01049,  7: 0.33181,  8: 0.25147,  9: 0.13661,
       |      | 10: 0.46918, 11: 0.21771, 12:-0.77177, 13: 0.85803, 14: 0.36066, 15: 0.24100, 16:-0.11801, 17: 0.38035, 18:-0.58954
  31   |      |  1:-0.04540,  2:-0.17208,  3:-0.75326,  4:-0.67854,  5: 0.11723,  6: 0.14634,  7: 0.46854,  8: 0.25567,  9: 0.10273,
       |      | 10: 0.03856, 11: 0.72777, 12: 0.09178, 13:-0.17763, 14: 0.88784, 15: 0.32567, 16: 0.09120, 17: 0.59603, 18: 0.61054
  32   |      |  1: 0.05964,  2: 0.93050,  3: 0.44125,  4: 0.59737,  5:-0.11717,  6: 0.46371,  7: 0.41048,  8:-0.25946,  9:-0.78326,
       |      | 10:-0.95043, 11:-0.48875, 12:-0.56244, 13: 0.67528, 14:-0.36425, 15:-0.51300, 16:-0.71214, 17:-0.26373, 18:-0.96074
  33   |      |  1:-0.04523,  2: 0.29995,  3:-0.16696,  4: 0.97045,  5: 0.15074,  6:-0.07382,  7:-0.75571,  8: 0.38204,  9: 0.56957,
       |      | 10:-0.66758, 11:-0.42599, 12: 0.74032, 13: 0.27330, 14: 0.16896, 15:-0.91725, 16:-0.23350, 17:-0.63018, 18: 0.18224
  34   |      |  1:-0.01409,  2:-0.10714,  3:-0.03456,  4: 0.08037,  5: 0.03032,  6:-0.00664,  7: 0.10846,  8: 0.05370,  9: 0.57844,
       |      | 10: 0.13721, 11:-0.10508, 12:-0.27687, 13: 0.06822, 14: 0.16896, 15:-0.21786, 16: 0.13417, 17: 0.05892, 18: 0.26530
  35   |      |  1:-0.16600,  2: 0.01081,  3:-0.46619,  4: 0.40902,  5:-0.17115,  6:-0.11116,  7:-0.50032,  8: 0.47854,  9: 0.02057,
       |      | 10:-0.81072, 11:-0.48623, 12:-0.45042, 13:-0.91713, 14: 0.16896, 15:-0.88688, 16:-0.15087, 17: 0.95018, 18:-0.55462
  36   |      |  1:-0.02784,  2: 0.50258,  3:-0.10733,  4: 0.67760,  5:-0.53320,  6: 0.19996,  7: 0.13760,  8: 0.45647,  9:-0.46205,
       |      | 10: 0.12567, 11:-0.53722, 12: 0.08932, 13: 0.29222, 14:-0.76894, 15: 0.53807, 16: 0.31724, 17: 0.19801, 18: 0.06151
  37   |      |  1:-0.09471,  2:-0.43343,  3:-0.88244,  4: 0.69351,  5:-0.14311,  6: 0.03679,  7: 0.56690,  8: 0.63869,  9: 0.14321,
       |      | 10: 0.54624, 11: 0.53164, 12: 0.80298, 13: 0.56362, 14:-0.76963, 15:-0.53270, 16: 0.34904, 17: 0.13081, 18:-0.92344
  38   |      |  1: 0.07310,  2: 0.60830,  3: 0.49011,  4:-0.57653,  5:-0.07363,  6:-0.06094,  7:-0.54325,  8:-0.03619,  9:-0.32424,
       |      | 10: 0.75680, 11: 0.39768, 12: 0.93678, 13: 0.87166, 14:-0.08198, 15: 0.80654, 16: 0.83811, 17: 0.45757, 18:-0.35069
  39   |      |  1:-0.17755,  2:-0.90173,  3: 0.17058,  4: 0.66256,  5: 0.31263,  6:-0.13870,  7: 0.92213,  8:-0.35069,  9: 0.24929,
       |      | 10: 0.69737, 11: 0.47870, 12:-0.35768, 13:-0.35768, 14: 0.31263, 15: 0.61162, 16: 0.92213, 17:-0.87259, 18: 0.24929
```

Figure E.6. Values of the weights between the units of the DHW NN Model

198

```
40 ─ | 1:-0.05918, 2:-0.30377, 3:-0.32231, 4:-0.26905, 5: 0.01007, 6:-0.01752, 7: 0.63183, 8: 0.36870, 9:-0.47679,
       10: 0.39646, 11:-0.34644, 12:-0.66325, 13:-0.61132, 14: 0.84034, 15: 0.82453, 16: 0.09226, 17:-0.47742, 18: 0.71849

41 ─ | 1:-0.36677, 2:-0.55183, 3:-0.23530, 4:-0.23142, 5:-0.25406, 6: 0.29214, 7:-0.42883, 8:-0.02391, 9: 0.15230,
       10: 0.38122, 11:-0.86089, 12: 0.08530, 13:-0.03437, 14:-0.43013, 15: 0.51137, 16:-0.59705, 17:-0.74224, 18: 1.03892

42 ─ | 1: 0.01382, 2:-0.32070, 3:-0.00058, 4:-0.67609, 5: 0.08852, 6: 0.14814, 7:-0.93654, 8:-0.07380, 9:-0.39471,
       10:-0.88520, 11:-0.36832, 12: 0.39940, 13:-0.40066, 14:-0.78474, 15:-0.32835, 16:-0.85231, 17:-0.10543, 18:-0.90757

43 ─ | 1: 0.12503, 2:-0.10133, 3:-0.59641, 4: 0.36127, 5:-0.58352, 6: 0.44495, 7:-0.83216, 8: 0.54588, 9:-0.19788,
       10:-0.06697, 11: 0.92756, 12: 0.26211, 13: 0.29045, 14:-0.40691, 15:-0.12102, 16:-0.30933, 17:-0.18167, 18: 0.37799

44 ─ | 1: 0.07678, 2: 0.00489, 3:-0.76424, 4: 0.04937, 5:-0.25507, 6: 0.15135, 7:-0.04464, 8:-0.18729, 9: 0.38505,
       10:-0.89608, 11: 0.01878, 12: 0.64647, 13:-0.02888, 14: 0.58348, 15: 0.46298, 16: 0.38541, 17: 0.70496, 18:-0.26756

45 ─ | 1: 0.00167, 2:-0.22509, 3: 0.57464, 4: 0.11430, 5:-0.43486, 6:-0.01227, 7: 0.69953, 8:-0.17112, 9:-0.11014,
       10: 0.62037, 11:-0.47922, 12:-0.38955, 13: 0.57793, 14:-0.42066, 15: 0.51943, 16: 0.49298, 17:-0.29197, 18:-0.76927

46 ─ | 1:-0.14484, 2:-0.56610, 3:-0.86784, 4:-0.16138, 5:-0.31463, 6: 0.34710, 7:-0.32631, 8: 0.72296, 9: 0.58302,
       10: 0.12411, 11: 0.21269, 12:-0.05676, 13:-0.66686, 14: 0.59571, 15:-0.08450, 16:-0.34814, 17: 0.74952, 18:-0.63630

47 ─ | 1:-0.16686, 2: 0.21882, 3: 0.16983, 4:-0.61394, 5: 0.40459, 6: 0.05335, 7:-0.04536, 8: 0.93614, 9:-0.51899,
       10: 0.59133, 11:-0.90171, 12: 0.49686, 13: 0.05978, 14:-0.62373, 15: 0.04263, 16: 0.29207, 17: 0.66764, 18:-0.42702
       19: 0.32460, 20:-0.26449, 21:-0.89916, 22: 0.66745, 23:-0.59336, 24: 0.88384, 25:-0.20296, 26:-0.35734, 27: 0.96770,

48 ─ | 28:-0.90402, 29: 0.30883, 30:-0.65401, 31:-0.17778, 32: 0.84249, 33:-0.73429, 34:-0.06025, 35: 0.50326, 36:-0.31571,
       37:-0.57910, 38: 0.90463, 39:-0.57913, 40:-0.20755, 41:-0.86420, 42: 0.19562, 43: 0.56730, 44: 0.33017, 45:-0.12479.
       46:-0.79710, 47:-0.85444.
```

Figure E.6. (continued) Values of the weights between the units of the DHW NN Model

The output of the DHW NN Model can be calculated by applying Equations E.5 and E.6. Since the hidden and output layer activation functions were chosen to be the logistic function (Equation 2.2), the outputs of the hidden and output layer units are calculated using the logistic function.

The output of each hidden layer unit is calculated as follows:

$$z_j = \frac{1}{1+e^{-\left(\sum_{i=1}^{18} x_i v_{ij} + b_j\right)}} \tag{E.5}$$

The output of the network is calculated as follows:

$$y_m = \frac{1}{1+e^{-\left(\sum_{j=19}^{47} z_j w_{jm} + b_m\right)}} \tag{E.6}$$

where,

x_i: input to the hidden layer unit j from input layer unit i

z_j: output of the hidden layer unit j / input to the output layer unit m from hidden layer unit j

v_{ij}:   weight between the input layer unit i and hidden layer unit j

w_{km}:   weight between the hidden layer unit j and output layer unit m

b_j: bias of the hidden layer unit j

b_m: bias of the output layer unit m

i: number of input layer units (1 to 18)

j: number of hidden layer units (19 to 47)

m: output layer unit (48)

3. SH NN MODEL

The NN model developed to estimate the SH energy consumption has 28 input layer units, two hidden layer, and one output layer unit. As shown in Figure E.4, the units from one to 28 represent the input layer units, units from 29 and 30 are the hidden layer units, and units 31 is the output layer unit.

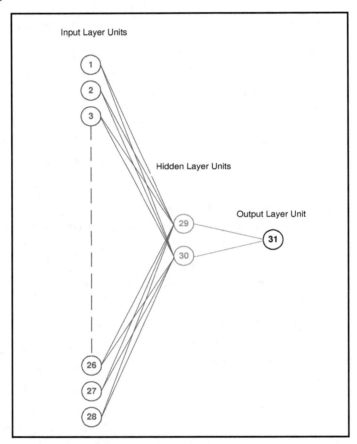

Figure E.7. Architectural configuration of the SH NN Model

The unit definition section of the SNNS output for the SH NN Model is given in Figure E.8. The bias values of the hidden and output layer units are given in the third column.

201

```
No. | Name                     | Bias      | Type   | Activation F. |
----|--------------------------|-----------|--------|---------------|
  1 | Dwelling Type            |  0.20707  | input  |               |
  2 | No. of Doors             | -0.68097  | input  |               |
  3 | No. of Triple Windows    |  0.29333  | input  |               |
  4 | No. of Double Windows    |  0.10683  | input  |               |
  5 | No. of Single Windows    |  0.36261  | input  |               |
  6 | Main Wall Area           | -0.77286  | input  |               |
  7 | Floor Area               |  0.96883  | input  |               |
  8 | Basement Wall Area       | -0.16818  | input  |               |
  9 | Basement Floor Area      |  0.68126  | input  |               |
 10 | Roof Area                |  0.23257  | input  |               |
 11 | Heated Basement          | -0.38201  | input  |               |
 12 | Heated Garage            |  0.88228  | input  |               |
 13 | Dwelling Age Cat.        | -0.79971  | input  |               |
 14 | Wall Age Cat.            |  0.67291  | input  |               |
 15 | Roof Age Cat.            |  0.68186  | input  |               |
 16 | Basement Wall Age Cat.   | -0.06613  | input  |               |
 17 | Basement Floor Age Cat.  | -0.20829  | input  |               |
 18 | System Efficiency        | -0.16540  | input  |               |
 19 | HRV                      |  0.46006  | input  |               |
 20 | Thermostats              |  0.21904  | input  |               |
 21 | Indoor Temperature       |  0.41320  | input  |               |
 22 | HDD                      | -0.48475  | input  |               |
 23 | Owner                    |  0.40295  | input  |               |
 24 | Income                   |  0.84892  | input  |               |
 25 | No. of Children          |  0.19003  | input  |               |
 26 | No. of Adults            | -0.71476  | input  |               |
 27 | Daytime Occupancy        | -0.88394  | input  |               |
 28 | Population               | -0.32566  | input  |               |
 29 |                          | -0.56311  | hidden | Identity      |
 30 |                          |  0.89065  | hidden | Identity      |
 31 |                          |  0.08948  | output | Logistic      |
----|--------------------------|-----------|--------|---------------|
```

Figure E.8. Unit definition of the SH NN Model

The values of the weights between the units of the SH NN Model are given in Figure E.9.

```
target | site | source:weight
-------|------|-------------
  29   |      | 1: 1.18914,  2: 0.26096,  3: 0.12264,  4: 0.43931,  5:-0.62584,  6: 0.38175,  7: 0.57426,  8:-0.01037,
       |      | 9:-0.01151, 10: 0.76295, 11: 0.10196, 12: 0.48369, 13:-1.20047, 14: 0.46053, 15: 0.35489, 16:-0.09730,
       |      |17:-0.32390, 18:-2.43354, 19: 0.13188, 20:-0.20061, 21: 0.13358, 22: 0.53347, 23:-0.68639, 24: 0.45175,
       |      |25: 0.15010, 26: 0.88421, 27:-0.38163, 28:-0.25620
-------|------|-------------
  30   |      | 1:-0.21833,  2: 0.54946,  3:-0.74418,  4: 0.84684,  5:-0.34325,  6: 0.36076,  7:-0.81657,  8:-0.03579,
       |      | 9:-0.45423, 10: 0.39038, 11:-0.47059, 12: 0.10650, 13:-0.11195, 14:-0.95793, 15:-0.59157, 16: 0.20185,
       |      |17: 0.49354, 18:-1.01231, 19: 0.36120, 20:-0.33685, 21:-0.02510, 22:-0.17016, 23:-0.68291, 24: 0.18199,
       |      |25:-0.51560, 26:-0.72227, 27: 0.23166, 28:-0.59323
-------|------|-------------
  31   |      |29: 0.36690, 30: 0.19933
-------|------|-------------
```

Figure E.9. Values of the weights between the units of the SH NN Model

The output of the SH NN Model can be calculated by applying Equations E.7 and E.8. Since the hidden layer activation function was chosen to be the identity function (Equation 2.4), the outputs of the hidden layer units are calculated using the identity function. Logistic function (Equation 2.2) was chosen for the output layer unit, thus the output of the unit at the output layer is calculated using the logistic function.

The output of each hidden layer unit is calculated as follows:

$$z_j = \sum_{i=1}^{28} x_i v_{ij} + b_j \tag{E.7}$$

The output of the network is calculated as follows:

$$y_m = \frac{1}{1 + e^{-\left(\sum_{j=29}^{30} z_j w_{jm} + b_m\right)}} \tag{E.8}$$

where,

x_i: input to the hidden layer unit j from input layer unit i

z_j: output of the hidden layer unit j / input to the output layer unit m from hidden layer unit j

v_{ij}:   weight between the input layer unit i and hidden layer unit j

w_{km}:   weight between the hidden layer unit j and output layer unit m

b_j: bias of the hidden layer unit j

b_m: bias of the output layer unit m

i: number of input layer units (1 to 28)

j: number of hidden layer units (29 to 30)

m: output layer unit (31)

APPENDIX F

CDA MODEL REGRESSION ANALYSIS

1. CDA ELECTRICITY MODEL

The SYSTAT commands for the regression analysis of the CDA EM are given in Figure F.1.

```
REGRESS

 USE "D:\Merih\Networks\CDA\ELECTRICITY\ELECTRICITY.SYD"

 LET WINDOW = TRIPLE + DOUBLE + SINGLE
 LET HHSIZE = CHILD + ADULT
 LET LIGHTS = HALO + FLOU + INCA

 LET PROGT1 = SH * PROGT
 LET HRV1 = SH * HRV
 LET AIT1 = SH * AIT
 LET DTYPE1 = SH * DTYPE
 LET AREA1 = SH * AREA
 LET AGECAT1 = SH * AGECAT
 LET BSMNT1 = SH * BSMNT
 LET GARAGE1 =  SH * GARAGE
 LET ATTIC1 = SH * ATTIC
 LET TRIPLE1 = SH * TRIPLE
 LET DOUBLE1 = SH * DOUBLE
 LET SINGLE1 = SH * SINGLE
 LET DOOR1 = SH * DOOR
 LET HDD1 = SH * HDD
 LET OWNER1 = SH * OWNER
 LET INCOME1 = SH * INCOME
 LET CHILD1 = SH * CHILD
 LET ADULT1 = SH * ADULT
 LET DAYTIME1 = SH * DAYTIME
 LET POPUL1 = SH * POPUL
 LET WINDOW1 = SH * WINDOW
 LET HHSIZE1 = SH * HHSIZE

 LET AREA2= SSH * AREA
 LET HDD2 = SSH * HDD
 LET AIT2 = SSH * AIT
 LET CHILD2 = SSH * CHILD
 LET ADULT2 = SSH * ADULT
 LET DAYTIME2 = SSH * DAYTIME
 LET HHSIZE2 = SSH * HHSIZE
```

Figure F.1 SYSTAT commands for the CDA EM regression analysis

```
LET TANK3 = DHW * TANK
LET SYSAGE3 = DHW * SYSAGE
LET BLANKET3 = DHW * BLANKET
LET PIPEINS3 = DHW * PIPEINS
LET LOWFLOW3 = DHW * LOWFLOW
LET AERATOR3 = DHW * AERATOR
LET GT3 = DHW * GT
LET CWLOAD3 = DHW * CWLOAD
LET DWLOAD3 = DHW * DWLOAD
LET DTYPE3 = DHW * DTYPE
LET OWNER3 = DHW * OWNER
LET INCOME3 = DHW * INCOME
LET ADULT3 = DHW * ADULT
LET CHILD3 = DHW * CHILD
LET HHSIZE3 = DHW * HHSIZE

LET CACUSE4 = CAC * CACUSE
LET DTYPE4 = CAC * DTYPE
LET AREA4 = CAC * AREA
LET AGECAT4 = CAC * AGECAT
LET ATTIC4 = CAC * ATTIC
LET TRIPLE4 = CAC * TRIPLE
LET DOUBLE4 = CAC * DOUBLE
LET SINGLE4 = CAC * SINGLE
LET DOOR4 = CAC * DOOR
LET CDD4 = CAC * CDD
LET OWNER4 = CAC * OWNER
LET INCOME4 =  CAC * INCOME
LET CHILD4 = CAC * CHILD
LET ADULT4 = CAC * ADULT
LET DAYTIME4 = CAC * DAYTIME
LET WINDOW4 = CAC * WINDOW
LET HHSIZE4 = CAC * HHSIZE

LET WACUSE5 = WAC * WACUSE
LET AREA5 = WAC * AREA
LET AGECAT5 = WAC * AGECAT
LET CDD5 = WAC * CDD
LET INCOME5 =  WAC * INCOME
LET CHILD5 = WAC * CHILD
LET ADULT5 = WAC * ADULT
LET DAYTIME5 = WAC * DAYTIME

LET INCOME6 = REF1 * INCOME
LET HHSIZE6 = REF1 * HHSIZE
LET INCOME7 = REF2 * INCOME
LET HHSIZE7 = REF2 * HHSIZE
```

Figure F.1 (continued) SYSTAT commands for the CDA EM regression analysis

```
LET INCOME8 = FREZ1 * INCOME
LET HHSIZE8 = FREZ1 * HHSIZE
LET INCOME9 = FREZ2 * INCOME
LET HHSIZE9 = FREZ2 * HHSIZE

LET HHSIZE10 = COOK * HHSIZE
LET MICROW2 = COOK * MICROW

LET AREA6 = FF * AREA
LET HDD6 = FF * HDD

LET AREA7 = BP * AREA
LET HDD7 = BP * HDD

LET HHSIZE11 = MICROW * HHSIZE
LET HHSIZE12 = CTV * HHSIZE
LET HHSIZE13 = BWTV * HHSIZE

LET HHSIZE14 = VCR * HHSIZE


PRINT=LONG

MODEL HEC = CONSTANT + SH + PROGT1 + HRV1 + AIT1 + DTYPE1 + AREA1 + ,
            AGECAT1 + BSMNT1 + GARAGE1 + ATTIC1 + TRIPLE1 + DOUBLE1 +,
            SINGLE1 + DOOR1 + HDD1 + OWNER1 + INCOME1 + CHILD1 + ,
            ADULT1 + DAYTIME1 + POPUL1 + ,
            SSH + AREA2 + HDD2 + AIT2 + CHILD2 + ADULT2 + DAYTIME2 +,
            DHW + TANK3 + SYSAGE3 + BLANKET3 + PIPEINS3 + LOWFLOW3 + ,
            AERATOR3 + GT3 + CWLOAD3 + DWLOAD3 + DTYPE3 + OWNER3 + ,
            INCOME3 + CHILD3 + ADULT3 + ,
            CAC + CACUSE4 + DTYPE4 + AREA4 + AGECAT4 + ATTIC4 + ,
            TRIPLE4 + DOUBLE4 + SINGLE4 + DOOR4 + CDD4 + OWNER4 + ,
            INCOME4 + CHILD4 + ADULT4 + DAYTIME4 + ,
            WAC + WACUSE5 + AREA5 + AGECAT5 + CDD5 + INCOME5 + ,
            CHILD5 + ADULT5 + DAYTIME5 + ,
            REF1 + VOLR1 + FROSTR1 + INCOME6 + HHSIZE6 + REF2 + ,
            VOLR2 + FROSTR2 + INCOME7 + HHSIZE7 + FREZ1 + VOLF1 + ,
            INCOME8 + HHSIZE8 + FREZ2 + VOLF2 + INCOME9 + HHSIZE9 +,
            COOK + HHSIZE10 + MICROW2 + DISH + DWLOAD + CLOTH + ,
            CWLOAD + DRYER + CDLOAD + FF + AREA6 + HDD6 + BP + ,
            AREA7 + HDD7 + MICROW + HHSIZE11 + CTV + HHSIZE12 + ,
            BWTV + HHSIZE13 + VCR + HHSIZE14 + HALO + FLOU + INCA
```

Figure F.1 (continued) SYSTAT commands for the CDA EM regression analysis

The results of the regression analysis of the CDA EM (Equation 4.28) are given in Figure F.2.

```
Dep Var: HEC    N: 2039   Multiple R: 0.813    Squared multiple R: 0.660

Adjusted squared multiple R: 0.657    Standard error of estimate: 5533.031

Effect      Coefficient    Std Error    Std Coef  Tolerance      t    P(2 Tail)

CONSTANT      2128.647      508.485       0.000        .       4.186    0.000
HDD1             1.595        0.177       0.392      0.089     9.036    0.000
HRV1         -2516.011     1077.603      -0.032      0.922    -2.335    0.020
DTYPE1        1891.746      722.629       0.085      0.160     2.618    0.009
AREA1           12.666        5.341       0.077      0.160     2.372    0.018
AGECAT1       -785.396      173.621      -0.133      0.196    -4.524    0.000
INCOME1         78.142       12.795       0.168      0.223     6.107    0.000
AREA2            8.126        2.473       0.051      0.694     3.286    0.001
CHILD2         569.330      248.449       0.037      0.659     2.292    0.022
TANK3           16.860        2.842       0.179      0.184     5.932    0.000
LOWFLOW3      -691.340      305.421      -0.034      0.732    -2.264    0.024
DWLOAD3        215.042       69.005       0.050      0.656     3.116    0.002
ADULT3         752.825      199.481       0.105      0.218     3.774    0.000
CACUSE4          1.274        0.737       0.024      0.887     1.728    0.084
FROSTR1       1030.230      357.785       0.038      0.953     2.879    0.004
FROSTR2       1636.477      579.324       0.043      0.714     2.825    0.005
INCOME7         28.235        7.695       0.060      0.634     3.669    0.000
VOLF1            1.495        0.611       0.033      0.928     2.447    0.014
HHSIZE10       421.249      106.811       0.065      0.620     3.944    0.000
CDLOAD         303.606       40.751       0.113      0.738     7.450    0.000
LIGHTS          50.175        8.876       0.088      0.694     5.653    0.000

                     Analysis of Variance
Source          Sum-of-Squares   df   Mean-Square      F-ratio       P
Regression        1.20051E+11    20   6.00253E+09     196.069     0.000
Residual          6.17799E+10  2018   3.06144E+07
```

Figure F.2. Regression analysis of the CDA EM

The eigenvalues of the X^TX matrix and the condition indices of the variables of the CDA EM are given in Figure F.3.

```
Eigenvalues of XᵀX

1            2            3            4            5
9.808        2.913        1.415        1.069        0.967

6            7            8            9            10
0.899        0.776        0.568        0.476        0.438

11           12           13           14           15
0.369        0.284        0.215        0.153        0.133

16           17           18           19           20
0.129        0.124        0.103        0.074        0.049

21
0.039


Condition indices

1            2            3            4            5
1.000        1.835        2.632        3.030        3.185

6            7            8            9            10
3.302        3.555        4.156        4.540        4.734

11           12           13           14           15
5.154        5.881        6.749        8.017        8.596

16           17           18           19           20
8.721        8.907        9.739        11.551       14.199

21
15.865
```

Figure F.3.   Eigenvalues of the X^TX matrix and the condition indices of the variables of the CDA EM

210

2. CDA NATURAL GAS MODEL

The SYSTAT commands for the regression analysis of the CDA NGM are given in Figure
F.4.

```
REGRESS

USE "D:\Merih\Networks\CDA\NATURALGAS\NATURALGAS.SYD"

LET WINDOW = TRIPLE + DOUBLE + SINGLE
LET HHSIZE = CHILD + ADULT
LET LIGHTS = HALO + FLOU + INCA

LET EFF1= SH * EFF
LET SHAGE1 = SH * SHAGE
LET PROGT1 = SH * PROGT
LET AIT1 = SH * AIT
LET DTYPE1 = SH * DTYPE
LET AREA1 = SH * AREA
LET AGECAT1 = SH * AGECAT
LET BSMNT1 = SH * BSMNT
LET GARAGE1 =  SH * GARAGE
LET ATTIC1 = SH * ATTIC
LET TRIPLE1 = SH * TRIPLE
LET DOUBLE1 = SH * DOUBLE
LET SINGLE1 = SH * SINGLE
LET DOOR1 = SH * DOOR
LET HDD1 = SH * HDD
LET OWNER1 = SH * OWNER
LET INCOME1 = SH * INCOME
LET CHILD1 = SH * CHILD
LET ADULT1 = SH * ADULT
LET DAYTIME1 = SH * DAYTIME
LET POPUL1 = SH * POPUL
LET WINDOW1 = SH * WINDOW
LET HHSIZE1 = SH * HHSIZE

LET AREA2= SSH * AREA
LET HDD2 = SSH * HDD
LET AIT2 = SSH * AIT
LET CHILD2 = SSH * CHILD
LET ADULT2 = SSH * ADULT
LET DAYTIME2 = SSH * DAYTIME
LET HHSIZE2 = SSH * HHSIZE
```

Figure F.4. SYSTAT commands for the CDA NGM regression analysis

```
LET TANK3 = DHW * TANK
LET SYSAGE3 = DHW * SYSAGE
LET BLANKET3 = DHW * BLANKET
LET PIPEINS3 = DHW * PIPEINS
LET LOWFLOW3 = DHW * LOWFLOW
LET AERATOR3 = DHW * AERATOR
LET GT3 = DHW * GT
LET CWLOAD3 = DHW * CWLOAD
LET DWLOAD3 = DHW * DWLOAD
LET DTYPE3 = DHW * DTYPE
LET OWNER3 = DHW * OWNER
LET INCOME3 = DHW * INCOME
LET ADULT3 = DHW * ADULT
LET CHILD3 = DHW * CHILD
LET HHSIZE3 = DHW * HHSIZE

LET HHSIZE4 = COOK * HHSIZE
LET MICROW4 = COOK * MICROW
LET CDLOAD5 = DRYER * CDLOAD
LET INCOME6 = POOL * INCOME

PRINT=LONG


MODEL HEC = SH + EFF1 + SHAGE1 + PROGT1 + AIT1 + DTYPE1 + AREA1 + ,
            AGECAT1 + BSMNT1 + GARAGE1 + ATTIC + TRIPLE1 + DOUBLE1 + ,
            SINGLE1 + DOOR1 + HDD1 + OWNER1 + INCOME1 + CHILD1 + ,
            ADULT1 + DAYTIME1 + POPULA1 +,
            SSH + AIT2 + AREA2 + HDD2 + CHILD2 + ADULT2 + DAYTIME2 + ,
            DHW + TANK3 + SYSAGE3 + BLANKET3 + PIPEINS3 + LOWFLOW3 + ,
            AERATOR3 + GT3 + CWLOAD3 + DWLOAD3 + DTYPE3 + OWNER3 + ,
            INCOME3 + CHILD3 + ADULT3 +,
            COOK + HHSIZE4 + MICROW4 + DRYER + CDLOAD5 +,
            POOL + INCOME6

ESTIMATE
```

Figure F.4. (continued) SYSTAT commands for the CDA NGM regression analysis

The results of the regression analysis of the CDA NGM (Equation 4.36) are given in Figure F.5.

```
Dep Var: HEC   N: 1003   Multiple R: 0.961   Squared multiple R: 0.923

Adjusted squared multiple R: 0.923   Standard error of estimate: 984.001

Effect       Coefficient    Std Error    Std Coef Tolerance     t     P(2 Tail)

PROGT1        -207.752        79.078       -0.026    0.778    -2.627    0.009
DOOR1          129.839        30.318        0.097    0.150     4.283    0.000
WINDOW1         26.302         5.505        0.102    0.168     4.778    0.000
SHAGE1          46.033         4.063        0.195    0.261    11.330    0.000
GARAGE1        437.688       111.559        0.037    0.884     3.923    0.000
AREA1            6.614         0.743        0.238    0.108     8.903    0.000
ADULT1         181.605        40.264        0.121    0.108     4.510    0.000
SYSAGE3         13.209         6.088        0.033    0.326     2.170    0.030
HHSIZE3         70.774        26.693        0.062    0.141     2.651    0.008
DTYPE3         720.967        85.233        0.184    0.164     8.459    0.000
HHSIZE4         94.039        53.298        0.016    0.919     1.764    0.078
CDLOAD5         35.910        21.770        0.015    0.903     1.650    0.099
POOL           969.555       377.900        0.023    0.969     2.566    0.010


                        Analysis of Variance

Source          Sum-of-Squares    df   Mean-Square      F-ratio        P

Regression       1.15628E+10      13   8.89443E+08      918.600      0.000
Residual         9.58576E+08     990   968258.515
```

Figure F.5. Regression analysis of CDA NGM

The eigenvalues of the X^TX matrix and the condition indices of the variables of the CDA NGM are given in Figure F.6.

```
Eigenvalues of unit scaled X^T X

1           2           3           4           5
7.073       1.164       1.020       0.878       0.785

6           7           8           9           10
0.744       0.407       0.282       0.205        0.161

11          12          13
0.124       0.091       0.067

Condition indices

1           2           3           4           5
1.000       2.465       2.634       2.839       3.002

6           7           8           9           10
3.083       4.170       5.007       5.873        6.638

11          12          13
7.558       8.819       10.270
```

Figure F.6.   Eigenvalues of the X^TX matrix and the condition indices of the variables of the CDA NGM

3. CDA OIL MODEL

The SYSTAT commands for the regression analysis of the CDA OM are given in Figure F.7.

```
REGRESS

USE "D:\Merih\Networks\CDA\OIL\OIL.SYD"

LET WINDOW = TRIPLE + DOUBLE + SINGLE
LET HHSIZE = CHILD + ADULT
LET LIGHTS = HALO + FLOU + INCA

LET EFF1= SH * EFF
LET SHAGE1 = SH * SHAGE
LET PROGT1 = SH * PROGT
LET AIT1 = SH * AIT
LET DTYPE1 = SH * DTYPE
LET AREA1 = SH * AREA
LET AGECAT1 = SH * AGECAT
LET BSMNT1 = SH * BSMNT
LET GARAGE1 =  SH * GARAGE
LET ATTIC1 = SH * ATTIC
LET TRIPLE1 = SH * TRIPLE
LET DOUBLE1 = SH * DOUBLE
LET SINGLE1 = SH * SINGLE
LET DOOR1 = SH * DOOR
LET HDD1 = SH * HDD
LET OWNER1 = SH * OWNER
LET INCOME1 = SH * INCOME
LET CHILD1 = SH * CHILD
LET ADULT1 = SH * ADULT
LET DAYTIME1 = SH * DAYTIME
LET POPUL1 = SH * POPUL
LET WINDOW1 = SH * WINDOW
LET HHSIZE1 = SH * HHSIZE


LET TANK3 = DHW * TANK
LET SYSAGE3 = DHW * SYSAGE
LET BLANKET3 = DHW * BLANKET
LET PIPEINS3 = DHW * PIPEINS
LET LOWFLOW3 = DHW * LOWFLOW
LET AERATOR3 = DHW * AERATOR
LET GT3 = DHW * GT
LET CWLOAD3 = DHW * CWLOAD
LET DWLOAD3 = DHW * DWLOAD
LET DTYPE3 = DHW * DTYPE
LET OWNER3 = DHW * OWNER
LET INCOME3 = DHW * INCOME
LET ADULT3 = DHW * ADULT
LET CHILD3 = DHW * CHILD
LET HHSIZE3 = DHW * HHSIZE
```

Figure F.7. SYSTAT commands for the CDA OM regression analysis

```
PRINT=LONG

MODEL HEC = SH + EFF1 + SHAGE1 + PROGT1 + AIT1 + DTYPE1 + AREA1 + ,
            AGECAT1 + BSMNT1 + GARAGE1 + ATTIC + TRIPLE1 + DOUBLE1 + ,
            SINGLE1 + DOOR1 + HDD1 + OWNER1 + INCOME1 + CHILD1 + ,
            ADULT1 + DAYTIME1 + POPULA1 +,
            DHW + TANK3 + SYSAGE3 + BLANKET3 + PIPEINS3 + LOWFLOW3 + ,
            AERATOR3 + GT3 + CWLOAD3 + DWLOAD3 + DTYPE3 + OWNER3 + ,
            INCOME3 + CHILD3 + ADULT3

ESTIMATE
```

Figure F.7. (continued) SYSTAT commands for the CDA OM regression analysis

The results of the regression analysis of the CDA OM (Equation 4.40) are given in Figure
F.8.

```
Dep Var: HEC   N: 231   Multiple R: 0.935   Squared multiple R: 0.875

Adjusted squared multiple R: 0.872   Standard error of estimate: 1095.544

Effect      Coefficient    Std Error    Std Coef Tolerance    t    P(2 Tail)

SHAGE1         40.046        8.384        0.206    0.300    4.777    0.000
AREA1           4.949        1.265        0.217    0.181    3.911    0.000
WINDOW1        41.681       10.282        0.205    0.219    4.054    0.000
DTYPE1        471.210      190.320        0.145    0.162    2.476    0.014
TANK3           5.277        0.814        0.246    0.388    6.480    0.000
DWLOAD3        50.907       29.073        0.050    0.682    1.751    0.081


                        Analysis of Variance

Source          Sum-of-Squares   df   Mean-Square    F-ratio      P

Regression       1.88179E+09      6   3.13632E+08    261.313    0.000
Residual         2.70049E+08    225   1200217.528
```

Figure F.8. Regression analysis of the CDA OM

The eigenvalues of the X^TX matrix and the condition indices of the variables of the CDA Oil Model are given in Figure F.9.

```
Model contains no constant

Eigenvalues of unit scaled XᵀX

1          2          3          4          5
4.318      0.822      0.340      0.259      0.139

6
0.122

Condition indices

1          2          3          4          5
1.000      2.292      3.564      4.081      5.568

6
5.946
```

Figure F.9. Eigenvalues of the X^TX matrix and the condition indices of the variables of the CDA OM

APPENDIX G

DERIVATIONS FOR THE COEFFICIENTS OF THE CDA MODEL VARIABLES

UNBIASED COEFFICIENTS FOR THE VARIABLES

The multivariable regression model can be shown in matrix notation as follows (Johnston and DiNardo, 1997):

$$y = X \beta + u \tag{G.1}$$

where,

y: dependent variable $n \times 1$ vector

X: independent variables $n \times k$ vector, the first column is for the intercept and is a column of ones

β: coefficients of the variables $k \times 1$ vector

u: disturbance $n \times 1$ vector

n: number of data points

k: number of independent variables

When the vector β is unknown, it is replaced by estimate b, and the unknown residual vector e is defined as

$$e = y - X b \tag{G.2}$$

The least square principle is used to determine b so the SSE (*i.e.* $e^T e$) is minimized:

$$
\begin{aligned}
SSE &= e^T e \\
&= (y - X b)^T (y - X b) \\
&= y^T y - b^T X^T y - y^T X b + b^T X^T X b \\
&= y^T y + 2 b^T X^T y + b^T X^T X b
\end{aligned}
\tag{G.3}
$$

The first order derivation of Equation G.3 with respect to b is as follows:

$$\frac{\partial(SSE)}{\partial b} = -2X^T y + 2X^T Xb = 0$$

which gives the below so called "normal equation":

$$(X^T X)\, b = X^T y$$

$$b = (X^T X)^{-1} X^T y \tag{G.4}$$

Substituting Equation G.1 into Equation G.4 gives:

$$
\begin{aligned}
b &= (X^T X)^{-1} X^T (X \beta + u) \\
&= \beta + (X^T X)^{-1} X^T u
\end{aligned}
\tag{G.5}
$$

Taking the expectations of the variables in Equation G.5 would give $E(b) = \beta$. This indicates that the coefficients of the variables are unbiased regardless of the multicollinearity problem.

Wissenschaftlicher Buchverlag bietet

kostenfreie

Publikation

von

wissenschaftlichen Arbeiten

Diplomarbeiten, Magisterarbeiten, Master und Bachelor Theses
sowie Dissertationen, Habilitationen und wissenschaftliche Monographien

Sie verfügen über eine wissenschaftliche Abschlußarbeit zu aktuellen oder zeitlosen
Fragestellungen, die hohen inhaltlichen und formalen Ansprüchen genügt,
und haben **Interesse an einer honorarvergüteten Publikation**?

Dann senden Sie bitte erste Informationen über Ihre Arbeit per Email
an info@vdm-verlag.de. Unser Außenlektorat meldet sich umgehend bei Ihnen.

VDM Verlag Dr. Müller Aktiengesellschaft & Co. KG
Dudweiler Landstraße 125a
D - 66123 Saarbrücken

www.vdm-verlag.de